Aerogels II
Preparation, Properties and Applications

Edited by

Inamuddin[1], Rizwana Mobin[2], Mohd Imran Ahamed[3] and Tariq Altalhi[4]

[1]Department of Applied Chemistry, Zakir Husain College of Engineering and Technology, Faculty of Engineering and Technology, Aligarh Muslim University, Aligarh-202002, India

[2] Department of Industrial Chemistry, Govt. College for Women, Cluster University, Srinagar, Jammuand Kashmir-190006, India

[3] Department of Chemistry, Faculty of Science, Aligarh Muslim University, Aligarh-202002, India

[4]Department of Chemistry, Collage of Science, Taif University, P.O. Box 11099, Taif 21944, Saudi Arabia

Copyright © 2021 by the authors

Published by **Materials Research Forum LLC**
Millersville, PA 17551, USA

Published as part of the book series
Materials Research Foundations
Volume 97 (2021)
ISSN 2471-8890 (Print)
ISSN 2471-8904 (Online)

Print ISBN 978-1-64490-128-1
eBook ISBN 978-1-64490-129-8

Distributed worldwide by

Materials Research Forum LLC
105 Springdale Lane
Millersville, PA 17551
USA
https://www.mrforum.com

Manufactured in the United States of America
10 9 8 7 6 5 4 3 2 1

Table of Contents

Preface

Aerogels are extremely porous and hence, lightest solid materials with huge surface areas and nanoscale pore sizes mainly in the micro- and mesoporous regimes. They are made of polymer with a solvent to form a gel, and then removing the liquid from the wet gels and replacing it with air without collapsing the 3D network structure. These unique characteristics make them promising materials which can be used in many applications including energy storage, thermal storage, catalysis, water splitting, environmental remediation, and among others. Therefore, to gather viewpoints and opportunities of aerogels are needed in the materials research community.

This book exclusively focuses on the properties and applications of aerogels. The chapters examine the synthetic methodologies, characterization tools, various types of organic, inorganic and hybrid aerogels, composites, and its applications in energy and environmental science. Written by leading experts in this field, this book targets scientists, faculty, and postgraduate students working with aerogels and their potential applications and should be of interest to readers working in the areas of chemistry, physics, polymer science, nanotechnology, and material science.

Key features:

- Gives a detailed account of aerogels properties and their applications
- Covering the basic concepts, methodologies, properties, and problems associated with aerogels
- Fits the background of various science and engineering disciplines
- Provides cutting-edge advances for aerogel technologies

Summary

Chapter 1 presents the latest findings on the development of polymeric aerogels, drying techniques, properties, and pharmacological applications. The functional properties such as biodegradability, low toxicity, and biocompatibility with the cellular matrix are addressed. Therefore, the production of polymeric aerogels is strategic and essential for the development of new products.

Chapter 2 discusses the diverse aerogel materials used across the globe for different biomedical applications. The recent applications of aerogels in drug delivery, implantable devices, regenerative medicine encompassing tissue engineering and bone regeneration, and biosensing are detailed in brief.

Chapter 3 focuses on the most recent discoveries about bioaerogels, addressing the synthesis, impregnation of bioactive compounds, pharmacological applications, aspects of cell uptake, biodegradability, and toxicity. Also, the production of molecular scaffolding, capsules for drug delivery, and development of biomaterials with properties compatible with the cellular environment were discussed.

In Chapter 4 a novel class of anisotropic, mechanically robust, and lightweight aerogel with excellent thermal insulation performance has been discussed. This chapter focuses on the thermal insulation, there is no second thought that the aerogel shows its superiority while making it a hybrid composite material.

Chapter 5 reviews about various carbon-based aerogels as electrocatalyst that has been employed to improve the methanol oxidation reaction and oxygen reduction reaction for direct methanol fuel cells and polymer electrolyte membrane fuel cells. Additionally, the effects of doping aerogels with graphene to improve the catalytic activity have been discussed.

Chapter 6 describes the fundamental aspects of different types of batteries and signifies the importance of aerogel for battery application. Moreover, a detailed description of the synthesis and application of a few major aerogel based electrodes reported in the literature was summarized. The advantages, limitations and its ability to meet the requirement were addressed.

Chapter 7 comprises various scientific strategies to emphasize the usage of aerogels to manage the energy storage system and fulfill the demand of energy. Additionally, the chapter focuses on thermal energy which is used in the multidisciplinary areas like in the removal of toxic contaminants, environmental remediations, and water purification.

Chapter 8 summarizes the classification and physicochemical properties of aerogels. This chapter also includes the recent challenges and some perspectives for high-performance aerogel-based sensor development in gas, water, pressure, strain and stress, hydrogen peroxide, and electrochemical sensors.

Chapter 9 discusses the application of aerogels as pesticides to manage different pests. Various aerogels-based pesticides are used to manage stored grain insect and veterinary pests. They are also used as carriers for active ingredients of pesticide formulations. So, aerogel-based pesticides could be a good alternative to avoid pesticide resistance development in pests.

Editors

Inamuddin[1], Rizwana Mobin[2], Mohd Imran Ahamed[3] and Tariq Altalhi[4]

[1]Department of Applied Chemistry, Zakir Husain College of Engineering and Technology, Faculty of Engineering and Technology, Aligarh Muslim University, Aligarh-202002, India

[2]Department of Industrial Chemistry, Govt. College for Women, Cluster University, Srinagar, Jammuand Kashmir-190006, India

[3] Department of Chemistry, Faculty of Science, Aligarh Muslim University, Aligarh-202002, India

[4]Department of Chemistry, Collage of Science, Taif University, P.O. Box 11099, Taif 21944, Saudi Arabia

Materials Research Forum LLC
https://doi.org/10.21741/9781644901298-1

Chapter 1

Polymer Aerogels: Preparation and Potential for Biomedical Application

Jhonatas Rodrigues Barbosa[1,*], Luiza Helena da Silva Martins[2],
Raul Nunes de Carvalho Junior[1,3]

[1]LABEX/FEA (Faculty of Food Engineering), Program of Graduation in Food Science and Technology, Federal University of Pará, Rua Augusto Corrêa S/N, Guamá, 66075-900 Belém, Pará, Brazil

[2]Institute of Animal Health and Production - Federal Rural University of Amazonia - Avenida Presidente Tancredo NevesN° 2501; Terra Firme; Cep: 66.077-830Belém, Pará, Brazil

[3]Program of Graduation in Natural Resources Engineering, Federal University of Pará, Rua Augusto Corrêa S/N, Guamá, 66075-900 Belém, Pará, Brazil

jhonquimbarbosa@gmail.com; Orcid iD: https://orcid.org/0000-0002-6394-299X.

Abstract

Polymeric aerogels have high added value and application. The potential to use natural polysaccharides, especially those from waste, has contributed to adding economic, social and ecological value. This chapter seeks to put forth the latest findings on the development of polymeric aerogels, drying techniques, properties, and pharmacological applications. The functional properties of polymeric aerogels, such as biodegradability, low toxicity and biocompatibility with cellular media are addressed. In the last decade, several works have reported the production of polymeric aerogels from natural polysaccharides. Chemical modifications and filling of new molecules were studied, improving the physicochemical and functional properties of the aerogels, as well as the drying techniques, were reported, and discarded. The production of polymeric aerogels is considered strategic for the development of sustainable, biodegradable and economically viable products.

Keywords

Polysaccharides, Supercritical Drying, Biodegradability, Nano Pores, Biocompatibility, Aerogels

Contents

1.   Introduction

Polymeric aerogels are porous solids from polysaccharides, the liquid part of which has been replaced by gases, are extremely light and pliable. The definition of aerogels has undergone revisions. According to Vareda, Lamy-Mendes, and Durães [1], the definition of aerogels should be revised, in light of recent technological advances, such as including drying methods. Based on recent discoveries, drying techniques, and studies on aerogels modifications and applications, this chapter focuses on the view that polymeric aerogels can be redefined as entangled polymeric matrices joined by complex chemical bonds having the following properties: high porosity, large surface area, pore distribution at the nano and micro scale and low density. Moreover, the ever-increasing application of natural polysaccharides in the formulation of polymeric aerogels has led the scientific community to a new definition that represents the sustainable base.

Over the past decade, we have seen the development of natural polysaccharide-based polymer aerogels. The production of natural polysaccharide-based polymeric aerogels was stimulated by the growing demand for products from sustainable processes. Suffice it to recall that conventional aerogels come from unsustainable resources such as inorganic and petrochemical materials. Based on these unsustainable resources, silica and graphene aerogels [2] and titanium aerogels and their oxides [3] were produced. Polysaccharide aerogels, especially those from natural resources, have excellent quality and physicochemical properties.

Polysaccharides from natural sources are fundamental bases for the production of polymeric aerogels. Among these polysaccharides, cellulose exists in abundance, widely distributed throughout the plant biomass and some microorganisms. The extraction of wood pulp, cotton, bamboo, banana fiber and residual biomass from agro-industrial processes has been explored [4,5]. In addition, other sources such as crustaceans and fungi have also been exploited to obtain polysaccharides that can be used as a basis for the production of polymeric aerogels. The biocompatibility, biodegradability, and abundance of some polysaccharides becomes key incredible for the production of new biomaterials. Due to the properties of polymeric earplugs, new products such as packaging foams have been manufactured, new bases for encapsulation, loading and controlled release of medicines, among others.

This chapter aims to report, discuss and compare recent findings in the literature on fabrication, drying, physicochemical properties, biomedical applications and potential of natural polysaccharides as bases for polymer aerogels production. Thus, the present chapter introduces the reader to the potential of various natural polysaccharides as interesting bases for aerogels production. Following the discussion on preparation, types of drying and finally biomedical applications of various types of polymer polymers based on natural polysaccharides.

2.   Polysaccharides as a basis for aerogels production

Polysaccharides are natural or artificially synthesized macromolecular biopolymers, of linear or branched chain and varying molecular weight [6]. These biopolymers are widely distributed in living organisms, where they play an important role in the biology and biochemistry of living organisms [7]. Polysaccharides are widely used in the production of polymeric aerogels because of their ability to form extensive intertwined networks, bio sustainability, biodegradability, biocompatibility, and innovative potential. Polysaccharides from natural sources are being widely explored from industry-resistant biomass to species in particular plants, fungi, algae, and bacteria. In this sense, the present topic will address some polysaccharides and their important characteristics for the production of polymeric aerogels.

2.1 Cellulose as the world's most abundant organic resource

Cellulose is present in the most varied organisms, mainly distributed mainly throughout the Plantae kingdom. Cellulose is a high molecular complex biopolymer composed of D-glucose and glycosidic bonds. This biopolymer is an extraordinary resource, abundant and recyclable, and can replace various synthetic polymers due to their properties, especially the mechanical strength and ability to form complex aerogels. Three main types of polymeric aerogels can be produced from cellulose, and are divided depending on the morphology, the production process, and the different ingredients. Thus, cellulose nanocrystalline aerogels, nanofibrillated cellulose aerogels and finally bacterial cellulose aerogels can be produced [8].

Cellulose nanocrystals are obtained by hydrolysis and mechanical treatment reactions, producing fibers, few nanometers in length [9]. The collapse of hydrogen bonds is prevented as demonstrated by Heath and Thielemans [10], producing high mechanical strength cellulose nanocrystals. Aerogels produced by cellulose nanocrystals, also known as nanowhiskers, are highly crystalline, have a highly rigid structure and high nanowhisker modulus. Cheng et al. [11] demonstrated that hydrophobic surface cellulose nanocrystals aerogels impregnated in cotton fabrics can be applied in solution partition processes.

A fibrillation process, i.e. production of cellulose fibers, or even microfibers that reach diameters from 5 to 70 nm [11], produces nanofibrillated cellulose. Cherian, Paulose and Vysakh [12], produced nanofibril cellulose polymeric aerogels with pores ranging from 10 to 100 nm after drying with supercritical fluid. Xiao et al. [13] reported the manufacture of nanofibril cellulose polymeric aerogels from pine needles. The small pore size is influenced by the high content of nanofibrilated cellulose, which hinders the growth of ice crystals and a mechanical dusting of the network.

Unlike nanofibrillated cellulose and cellulose nanocrystals, bacterial cellulose is produced through complex biochemical synthesis processes coming from the bacterial cell wall. The main microorganism responsible for bacterial cellulose production is *Acetobacter xylinum* bacterium, as explained in (Box.1), on bacterial cellulose production. Hosseini, Kokabi and Mousavi [14] reported the production of bacterial cellulose polymeric aerogels with tangled nanofiber networks. The addition of reduced graphene oxide nanocomposites to bacterial cellulose polymeric aerogels leads to research for applications in sensor production areas and even as components and parts. Due to biocompatibility, permeability to liquids and gases, high mechanical resistance, stimulating epithelialization capacity and inhibition of skin infections, bacterial cellulose polymeric aerogels can be used in the medical area, especially in skincare, drug delivery, and healing of wounds [15].

Box. 1| Biotechnology Connection |

Bacterial cellulose production as potential for synthesis of polymeric aerogels.

Bacterial cellulose is a high molecular biopolymer produced by certain types of bacteria. Bacterial cellulose production by microorganisms outside the cellular environment is a biology strategy in response to physicochemical changes in the extracellular environment. Under abnormal conditions of pH, nutrient composition, temperature, and oxygen flow, the bacteria are forced to produce extracellular polysaccharides for cell wall protection [16].

Various types of celluloses can be produced from biotechnological processes using different types of bacteria. In addition, different structures, morphologies, physicochemical, molecular properties and applications can be designed. Methods of producing cellulose in static, agitated and bioreactor cultures have been reported [17]. Cultivation submerged in static media results in the formation of cellulose films on the surface of the culture solution. On the other hand, production in agitated media, such as aerated bioreactors, bacterial cellulose production results in spheres, granules, and irregular mass. The type of cultivation technique chosen depends on the purpose of the application, the desired physical, morphological and mechanical characteristics [17].

Several culture parameters such as pH, culture medium composition, agitation, temperature, and oxygen flow affect the properties of bacterial cellulose. In recent years, several studies [18-20] have reported the feasibility of using agricultural residues as the basic carbon and nitrogen source for bacterial culture and bacterial cellulose production. In particular, bacterial cellulose production in agitated media such as bioreactors has developed, contributing to the expansion of large-scale production. Production in bioreactor systems is more efficient, controllable and scalable, facilitating system reproduction in industrial-scale biotechnological processes [20]. In (Fig. 1) we have the schematic representation of a bioreactor for bacterial cellulose production.

2.2   Plant polysaccharide main starch

Starch is a biopolymer with unique physicochemical and structural properties. The presence of starch in plants reveals much of the basic structural biology of these living organisms; in fact, starch is the main and most important storage polymer in plants. The presence of starch in plants partly explains the complexity of mechanisms involved in the cycle of embryonic growth, reproduction, and development [21]. Starch is a complex polymer formed of high molecular people glucose monomers consisting of two independent and interconnected units called amylose and amylopectin. Amylose is a long chain of unbound ($\alpha 1 \rightarrow 4$) linked-linked (β-glucan) residues. The molecular mass of this chemical structure ranges from thousands to millions of daltons. Amylopectin, as well as amylose, has a high molecular chemical structure, however, it has several branching points. Long (β-glucan) residues connected by bonds (α1 → 4) form the main structure of amylopectin. Whereas, the branches are of the type 1 → 6-β-glucan and occur frequently every 24 and 30 residues [21].

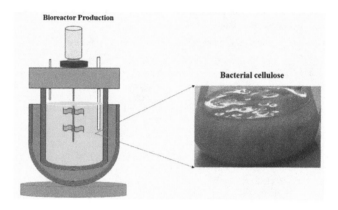

Fig. 1 *Bacterial cellulose production in bioreactor.*

The viscosity of aqueous starch solutions depends on the size of their chemical structure, composition, molecular people, the spatial shape of their molecules and the spatial conformation they may occupy. The greater the degree of freedom between glycosidic bonds, the greater the spatial possibilities of conformation. Due to the degree of freedom between the glycosidic bonds, a strong state of entropy is generated, overcoming the energetic conditions, inducing the chain in aqueous solution to adopt random disordered or helical spatial conformations [21].

Depending on the properties of starch, such as branched structures, high molecular weight and freedom between glycosidic bonds, three-dimensional, interlaced structures may be formed in aqueous solution. These three-dimensional structures can be used for polymer aerogel formulations. According to the starch composition in the percentage of amylose and pectin, various types of polymeric aerogels can be formed. According to Druel et al. [22], the starch composition is so relevant to the formation of polymer aerogel that the choice of high amylopectin starches should be prioritized.

According to Zhu [23], the addition of ideal proportions of amylose can increase the surface area of aerogels and reduce density. In addition, the composition of the starch and the proportion of its constituents influences the microstructure of the aerogels. Despite the qualities of starch as potential for application in the production of polymeric aerogels, there are some disadvantages such as low hardness and resistance to high pressures. Thus, some works such as Abhari, Madadlou, and Dini [24] and Miao et al. [25] reported on the modification of starch by the inclusion of specific chemical structures, such as trisodium citrate with strong network connection properties, improving the hardness and decreasing hardness the adhesiveness on the polymeric aerogels.

2.3   Other polysaccharides

In addition to the polymer matrices already reported, other polysaccharides such as chitosan, konjac glucomannan, alginate and pectin might be applied in the formulation of polymeric aerogels. Chitosan is the main deacetylated polysaccharide from arthropods and mushrooms. Chitin is the precursor polysaccharide of chitosan; chemical deacetylation processes are efficient in the recovery of chitosan fractions. Chitosan can be improved by adding chemical groups such as monomethyl [26], graft reaction, O-alkylated, among others [27]. In Hassan's research, Suzuki and El-Moneim [28], chitosan aerogel with three-dimensional porous scaffolding were designed and it was observed that the hierarchical structure was affected by electrochemical cyclability and rate capacity. Subsequently, Takeshita and Yoda [29] demonstrated by comparison that chitosan polymeric aerogels, relative to cellulose aerogels, have a more randomly oriented pore structure, contributing to thermal insulation properties.

Konjac glucomannan is a complex polysaccharide formed by β-1 $\rightarrow$ 4 bonds with D-mannose and D-glucose residues with acetyl substitution. The presence of several chemical groups ($-COOH$ and OH) in the skeletal unit is responsible for attracting multivalent cations, forming desirable chemical cross-linking structures [30]. In studies by Wang et al. [31], it has been shown that the addition of gelatin and starch to the Konjac glucomannan skeleton may help in the formation of open and closed pores. Moreover, studies such as those by Xin et al. [32] demonstrated that modifications to the basic structure of Konjac glucomannan could lead to interesting functions.

Alginate and pectin are other polysaccharides with interesting properties. As demonstrated by the work of Shao et al. [33], N-isopropyl acrylamide (hydrophobic block) and N-hydroxymethyl acrylamide (hydrophilic block) alginate polymeric aerogels have a thermal response and intelligent pH response. Pectin has been studied for its stabilizing properties, high viscosity and frying capacity of gels, as well as pectin polymeric aerogels, have been designed for the administration of gastro-resistant drugs

Materials Research Forum LLC

https://doi.org/10.21741/9781644901298-1

[34]. Modifications such as carboxymethylation, functional copolymerization and oxidation of the basic structure may improve the properties of aerogels [35].

3.   Preparation of polymeric aerogels

3.1   Polymeric aerogels can be designed to have defined pores

Polymeric aerogels may be designed to have defined pores, based on the properties of the aerogels, drying methods and polymer base type, may be synthesized aerogels with varying pores. Aerogel has advantages over other materials because the pores can be designed according to the purpose of the aerogels application. The porous channels of polymeric aerogels are advantageous due to the pore quality, size, thickness and low density of the polymer associated with the aerogels, which contribute to define the type of application [36].

Depending on the application requirement, polymeric aerogels may be designed to have open or close pores. In many situations, the pores are a limiting factor, influencing the functional properties and affecting the quality of a product. In this way, the pores of the aerogel can be designed to have the quality of communicating which would facilitate drug delivery, drug and biopharmaceutical encapsulation for example. In specific cases such as drug delivery, pores are an indispensable requirement. For example, in situations where aerogels are used in regenerative medicine, uniform and open-pore scaffolding structures are ideal. However, in cases where aerogels are applied as pollutant adsorbents, a mix of close and open pores is advisable. Therefore, the requirement of structuring defined pore-polymeric air-synthesis synthesis projects is strategic to meet application requirements [36].

Oschatz et al. [37] and Zhu et al. [38], the type of air preparation technology, the drying methods, and the polymeric raw material are the main factors influencing pore type and quality to be designed. Based on polymer chain type, functional groups, fillers, and molecular people, polymeric aerogels may have different pore structures. It is noteworthy that the synthesis approaches directly affect the properties of the aerogels. These conclusions are corroborated by the work of Wang et al. [31], who demonstrated that starch concentration influences pore type.

Cuadros, Erices, and Aguilera [39], observed that the addition of different polymers forming polymeric complexes might be an innovative strategy for producing defined pore polymeric aerogels. These authors evaluated the production of polymer-based aerogels from three different solutions. Infrared microscopy and spectroscopy results suggest that the addition of gelatin produces interconnected pores. Cuadros, Erices, and Aguilera [39] observed similar results and Pircher et al. [40], in which gelatin, sucrose, and paraffin were used as specific chemical connectors, forming the basis of well-structured pores.

However, the main and best-evaluated method for the production of structured pores is drying methods. Due to the quality and the principles of transport and mechanical turbulence associated with the gas flow, the drying methods are good for controlling the type of pore structure that will be developed. This topic will be covered in detail below. However, it should be noted that supercritical drying has been a very efficient method for producing nanoscale pores. Groult and Budtova [34], when studying various types of parameters on the impact of the external pore structures of aerogel approached this method. It has been shown that the drying method influences aerogel structures.

4.   Drying of polymeric aerogel

4.1   Supercritical drying

Supercritical fluid technology can be applied in the manufacture and drying of polymeric aerogel. There are several compounds that can be used as supercritical fluids, however carbon dioxide (CO_2) is the most used for several reasons, however the three main ones will be presented: green solvent, with low toxicity to humans and the environment; thermodynamic temperature conditions, and critical pressures are moderate ($Tc = 31$ °C and $Pc = 72.9$ bar). Finally, perhaps the most fundamental issue, the polymer is preserved from solvent contamination and air contact, which could cause or trigger oxidation reactions [41]. In (Fig. 2) a simple description of the supercritical drying plant and alcohol removal mechanism of the basic aerogel structure, in addition to the PT and PV phase diagrams for a pure substance is presented.

The curves that separate each of the regions are called the saturation curve and represent the equilibrium state between the phases. The vaporization curve, the melting curve and the sublimation curve can be observed. The point joining the three curves is called the triple point and represents the thermodynamic condition of coexistence between the three phases [41]. The vaporization curve ends at a thermodynamic point defined as a critical point, at which any pure substance from that point is in the supercritical state [42].

In the PV phase diagram, it is noted that isotherms under thermodynamic conditions of higher than critical temperature (Isotherm T> Tc) are a curve that does not cross the boundary between the phases. However, in the critical isotherm (T = Tc), the critical point is the inflection point of the curve, i.e., at this point the liquid and vapor phases become indistinguishable. This thermodynamic phenomenon characterized by a high rate of change in solvent density with pressure and temperature, because of this variation, there is a consistent increase in the isothermal compressibility coefficient. Small changes in pressure and temperature affect this thermodynamic phenomenon of a consistent increase in isothermal compressibility coefficient near the critical point. The result is solvent density variations, improving solvation efficiency, allowing greater selectivity and matrix penetration [42].

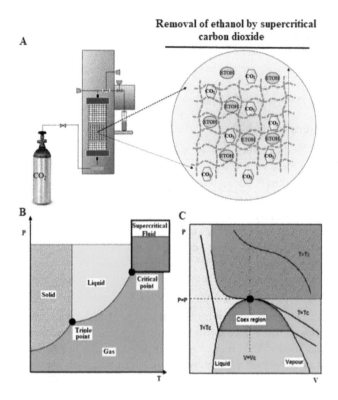

Fig. 2 *Supercritical fluid drying of polymeric aerogels. a) Simplified scheme showing the supercritical drying plant, with supercritical CO_2 pastures for alcohol removal from the basic aerogels structure, producing pores. b) Phase diagram for a pure substance (Pressure x Temperature). c) Phase diagram for a pure substance (Pressure x Volume).*

The characteristics of supercritical fluids such as relatively low viscosity and relatively high diffusivity make it easier for supercritical fluids to diffuse more easily through solid matrices and thus provide better results in mass and energy transfer efficiency between the matrix, improving extraction of natural compounds or facilitating nanoscale pore aberration in polymeric aerogel [43]. Another important feature is the ability to modify fluid density when pressure and temperature change.

The supercritical drying process is classified according to two approaches, low temperature, and high-temperature supercritical drying. When the drying method is chosen is the high-temperature one, the hydration water of the polymeric aerogels should be replaced by an organic solvent, then heated and pressurized [44]. When the solvent reaches the supercritical state it exits at constant temperature ventilation. Polymeric aerogels are dried as a function of the rapid expansion of the solvent and its volatilization when it reaches the supercritical state. When drying is done by the low-temperature method carbon dioxide is used as the drying medium. Thus, the organic solvents are replaced by supercritical carbon dioxide, which upon expansion expels the organic solvents in the microstructure of the aerogels. The main advantage of using supercritical solvents is related to the elimination of capillary pressure, which keeps the original structure of polymeric earrings intact [45].

4.2 Freeze-drying

The process of freeze-drying polymeric aerogel seems to be a current trend. The process itself is considered simple, however crystal growth and ice nucleation mechanisms are more complex. Freeze-drying is a drying process based on the principles of the sublimation of ice crystals. Thus, the sol-gel is previously frozen and then inside the lyophilizer goes through the change of physical state (solid to gas) sublimation process. Replacing ice crystals with gas under a high vacuum produces pores of various sizes [46]. The whole process of pore formation in polymeric aerogel that has undergone the freeze-drying process is determined by the growth phenomena of ice crystals in the sol-gel solution [46]. In (Fig. 3), a schematic representation depicting the process of water removal from aerogel by freeze-drying is shown.

As shown by Liu et al. [47], the formation of ice crystals during the manufacture of gum/graphene oxide aerogels exerts a force between the solutes. As demonstrated by the authors the force of the ice crystals pushes and assembles the hybrid solutes between the ice blocks, and then during sublimation, this same force is exerted on the wall of the formed aerogels, which in short confirms the hypothesis of the pore structure of polymeric aerogels is defined directly by the structure of ice crystals. Ni et al. [46], using a low temperature polarizing microscope, observed that the growth of ice crystals is influenced by the temperature and the solute concentration of the medium. Scanning electron microscopy (SEM) images confirmed that temperature and solution concentrate have a direct influence on the size and distribution of pores. The authors conclude that lyophilization can be used as an inexpensive and efficient technology for drying polymeric aerogels and yet the entire process of aerogels formation can be controlled to obtain defined pores.

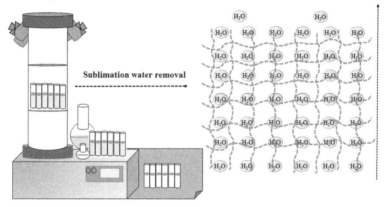

Fig. 3 *Scheme depicting the removal of water from polymeric aerogels by the freeze-drying method.*

Gupta et al. [48] synthesized low density, high mechanical strength nanofibrillated cellulose polymeric aerogels using the Freeze-drying method. Aerogels after field emission scanning electron microscopy analysis were characterized as porous (range 2 to 50 nm). Due to the properties of the nanofibrillated cellulose polymer aerogel, it would be an interesting candidate for applications as energy absorber and thermal insulator. Geng [49] addressed the manufacture and polymeric cellulose aerogels modified with N, N′-methylene bisacrylamide. The aerogel produced had thermal stability, macroporous structure, three-dimensional mesh, and absorption capacity of copper ions and methylene blue. In addition, different polymeric aerogel morphologies can be designed as a function of adjusting the degree of cellulose crosslinking.

5.   Biomedical applications

Polysaccharide-based polymeric aerogels may be used for various medical and pharmaceutical applications and procedures. In recent years the growing demand for more versatile, cheap, biodegradable, biocompatible and ecological bases for drug and drug loading has been investigated. In this context, polymeric aerogels have been explored. As shown in (Fig. 4), polymeric aerogels may be used for the delivery of drugs and bioactive compounds. In addition, modifications to the basic structures of polymeric aerogels can produce thermoresistant, intelligent and pH-change-resistant structures. All these properties are used in the pharmaceutical industry for the production of drugs that accelerate blood coagulation, which has antimicrobial activity among others. As will be shown in the next subtopics, chitosan, pectin, cellulose, and alginate polymeric aerogels have been used for various pharmacological and medical applications. It is important to remind the reader of the relevance of polysaccharide polymer bases to the production and modification of aerogels. Polysaccharide bases are considered versatile, as they have several functional groups in side chains available for chemical modification and molecular grafting procedures, which contributes to the application of these polymeric bases in the formulation of polymeric aerogels.

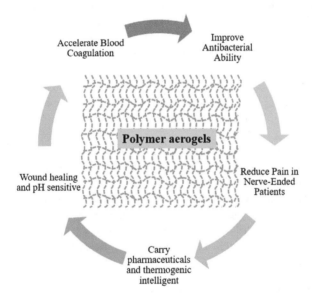

Fig. 4 *Pharmacological and medical applications of polymeric aerogels based on natural polysaccharides.*

5.1   Chitosan aerogels may accelerate blood coagulation

The wound healing process is quite complex, involving a collaborative set of various tissues, biochemical processes such as the release of antioxidant compounds, healing agents, and cell growth accelerators. Although natural cellular processes are efficient, various complications such as those associated with ulcers, chronic wounds, low immunity, and diabetes are challenging for the basic defense organism. Thus, to counteract these persistent inflammatory processes, aerogels can be used as carriers of potentially healing drugs. Chitosan polymeric aerogels have already been explored for their applicable medicinal properties. It has been shown that chitosan polymeric aerogels can accelerate sanguine coagulation in patients, reduce pain in patients with nerve endings, and still possess antimicrobial activity [50]. López-Iglesias et al. [51], using vancomycin-loaded chitosan aerogel, applied it to local ferrets, and obtained possible results for the reduction of infection and inflammation. While Duong et al. [52] combined the use of chitosan polymeric aerogels with cotton to repair and stop bleeding in stings. Chitosan aerogels have been shown to help neutralize and control systolic blood pressure at the edge of wounds, preventing seizure bleeding.

5.2   Pectin aerogel can carry pharmaceuticals

Pectin is a natural polysaccharide found mainly in the rind of some fruits such as cocoa and passion fruit. Due to biocompatibility, functionality, low toxicity and interesting functional groups for molecular couplings, pectin has been explored as a basis for drug grafting. In addition, some properties of these polymers such as efficient gelation, gel stability and high viscosity have caught the attention of researchers worldwide. Tkalec, Knez and Novak [53], explored the dissolution of some drugs in the pectin aerogel matrix with a high methoxyl load. The release of nifedipine coupled to the aerogel was 100% achieved within 12 hours, which demonstrates that these aerogel are interesting bases for drug release. In addition Horvat, Pantić, Knez and Novak [54], using an alginate and pectin-based hybrid shell-shell hybrid ear formulation, were able to prolong drug activity for a longer time.

5.3   Cellulose aerogel is used for wound healing

Cellulose is a natural polysaccharide present in plant matrices and some bacteria. It is a high molecular polymer and has an important structural function in plants. Currently, the main residues of agroindustry are the base of cellulose material, which has contributed to numerous studies such as the objective of reusing biomass. Some research has explored the potential of lignocellulosic biomass to extract and purify cellulose. Some applications with this matrix have already been developed, and the production of cellulose-based aerogels has already been a reality. Valo et al. [55], using cellulose as a base, produced high network complexity polymeric aerogels. The application of beclomethasone dipropionate nanoparticles to cellulose aerogels was efficient and positive results were achieved in wound healing. Wan and Li [56], using wheat straw as a source of silver cellulose, polypyrrole, and nanoparticles, synthesized cellulose aerogels with silver functional compounds, which have the potential to eliminate *Escherichia coli*, *Staphylococcus aureus* and *Listeria monocytogenes* bacterium direct contact.

5.4   Alginate aerogel is thermogenic intelligent and pH sensitive

Alginate polymer aerogels are negatively charged and can be used as a perfect fit with other oppositely charged polymers, forming versatile polyelectrolytic interactions. When alginate aerogel is combined with divalent cations, special egg-box structures are formed and drug delivery can be applied because the molecular structure is intelligent and sensitive to pH changes. Bugnone et al. [57] synthesized silica and alginate aerogels and evaluated the release rate of ketoprofen in water. Due to the sensitivity of the silica and alginate platform to pH, it was possible to simulate drug release under gastric conditions (pH, 1.2), as well as in the intestinal environment (pH, 6.8). The results showed that as a result of pH change the release rate of the drug was affected, indicating that aerogels platform can be used for intelligent drug release. Shao et al. [58], found that alginate aerogel with N-isopropylacrylamide and N-hydroxymethylacrylamide have double response to the release of hydrophobic drugs. The presence of two thermogenic compounds and the sensitivity to pH make this type of intelligent aerogels as an interesting possibility for the pharmaceutical industry. The presence of thermogenic compounds has two-way action; the first is the erosion of aerogel at low temperature and shrinkage at high temperature to the faster release kinetics of the drug when in neutral solution.

Conclusions

Polymeric aerogels are a new class of materials with high porosity, considered sustainable materials due to the use of natural polysaccharides and have unique properties such as biodegradability, low toxicity, and biocompatibility. Recent findings in the literature have revealed that natural polysaccharide-based polymeric aerogels have broad advantages compared to other types of aerogels. Due to these properties, polymeric aerogels have been applied in several sectors; however, the biomedical and pharmaceutical application has been highlighted. Based on the research reports, we understand that the current concept of polymeric aerogels should be updated considering the properties of nanopores and drying processes. Also, drying technologies, especially supercritical, offer interesting advantages in obtaining well-defined pore polymeric aerogels. Polymeric aerogels offer relative advantages of indirect drug delivery and controlled drug delivery. Finally, polymeric aerogels can be considered strategic technologies for the development of sustainable and biodegradable products.

Acknowledgment

The author's acknowledgment CNPq (National Council for Scientific and Technological Development), process number 169983/2018-8, scholarship grant and UFPA (Federal University of Pará), Brazil, for the space of development and scientific research, and all contributing authors.

References

[1] J.P. Vareda, A. Lamy-Mendes, L. Durães, A reconsideration on the definition of the term aerogel based on current drying trends, Micropor. Mesopor. Mat. 258 (2018) 211-216. https://doi.org/10.1016/j.micromeso.2017.09.016

[2] Y. Jiang, S. Chowdhury, R. Balasubramanian, New insights into the role of nitrogen-bonding configurations in enhancing the photocatalytic activity of nitrogen-doped graphene aerogels, J. Colloid Interface Sci. 534 (2019) 574-585. https://doi.org/10.1016/j.jcis.2018.09.064

[3] C. Zhang, S. Liu, Y. Qi, F. Cui, X. Yang, Conformal carbon coated TiO_2 aerogel as superior anode for lithium-ion batteries, Chem. Eng. J. 351 (2018) 825-831. https://doi.org/10.1016/j.cej.2018.06.125

[4] K. Harini, K. Ramya, M. Sukumar, Extraction of nano cellulose fibers from the banana peel and bract for production of acetyl and lauroyl cellulose, Carbohydr. Polym. 201 (2018) 329-339. https://doi.org/10.1016/j.carbpol.2018.08.081

[5] C.R. Bauli, D.B. Rocha, S.A. Oliveira, D.S. Rosa, Cellulose nanostructures from wood waste with low input consumption, J. Clean.Prod. 211 (2019) 408-416. https://doi.org/10.1016/j.jclepro.2018.11.099

[6] S.Z. Xie, R.H. Xue-Qiang, Z.L.H. Pan, J. Liu, J. P. Luo, Polysacchride of *Dendrobium huoshanense* activates macrophages via toll-like receptor 4-mediated signaling pathways, Carbohydr. Polym. 146 (2016) 292-300. https://doi.org/10.1016/j.carbpol.2016.03.059

[7] C.W. Cho, C.J. Han, Y.K. Rhee, Y.C. Lee, K.S. Shin, J.S. Shin, K.T. Lee, H. D. Hong, Cheonggukjang polysaccharides enhance immune activities and prevent

cyclophosphamide-induced immunosuppression, Int. J. Biol. Macromol. 72(2015) 519-525. https://doi.org/10.1016/j.ijbiomac.2014.09.010

[8] C. Wan, Y. Jiao, S. Wei, L. Zhang, Y. Wu, J. Li, Functional nanocomposites from sustainable regenerated cellulose aerogels: A review, Chem. Eng. J. 359 (2019) 459–475. https://doi.org/10.1016/j.cej.2018.11.115

[9] J.P. Oliveira , G.P. Bruni, S. L. M. el Halal, F. C. Bertoldi, A. R .G. Dias, E.R. Zavareze, Cellulose nanocrystals from rice and oat husks and their application in aerogels for food packaging, Int. J. Biol. Macromol. 124 (2019) 175-184. https://doi.org/10.1016/j.ijbiomac.2018.11.205

[10] L. Heath, W. Thielemans, Cellulose nanowhisker aerogels, Green Chem. 12 (2010) 1448–1453. https://doi.org/10.1039/c0gc00035c

[11]Q.Y. Cheng, C.S. Guan, M. Wang, Y.D. Li, J.B. Zeng, Cellulose nanocrystal coated cotton fabric with superhydrophobicity for efficient oil/water separation, Carbohydr. Polym. 199 (2018) 390–39. https://doi.org/10.1016/j.carbpol.2018.07.046

[12] J. Cherian, J. Paulose,P. Vysakh, Harnessing nature's hidden material: nano-cellulose. Mater. Today. 5 (2018) 12609–1261. https://doi.org/10.1016/j.matpr.2018.02.243

[13] S. Xiao, R. Gao, Y. Lu, J. Li, Q. Sun, Fabrication and characterization of nanofibrillated cellulose and its aerogels from natural pine needles. Carbohydr. Polym. 119 (2015) 202–209. https://doi.org/10.1016/j.carbpol.2014.11.041

[14] H. Hosseini, M. Kokabi, S.M. Mousavi, Conductive bacterial cellulose multiwall carbon nanotubes nanocomposite aerogel as a potentially flexible lightweight strain sensor. Carbohydr. Polym. 201 (2018) 228–235. https://doi.org/10.1016/j.carbpol.2018.08.054

[15] A. Sheikhia, J. Hayashi, J. Eichenbaum, M. Gutin, N. Kuntjoro, D. Khorsandi, A. Khademhosseini, Recent advances in nanoengineering cellulose for cargo delivery. J. Control. Release. 294 (2019) 53–76. https://doi.org/10.1016/j.jconrel.2018.11.024

[16] E.E. Kiziltas, A. Kiziltas, D. J. Gardner, Synthesis of bacterial cellulose using hot water extracted wood sugars. Carbohydr. Polym. 124 (2015) 131–138. https://doi.org/10.1016/j.carbpol.2015.01.036 0144-8617

[17] S.M. Santos, J.M. Carbajo, E. Quintana, D. Ibarra, N. Gomez, M. Ladero, M.E. Eugenio, J.C. Villar, Characterization of purified bacterial cellulose focused on its use on paper restoration. Carbohydr. Polym. 116 (2015) 173–181. https://doi.org/10.1016/j.carbpol.2014.03.064

[18] C. Campano,A. Balea, A. Blanco, C. Negro, Enhancement of the fermentation process and properties of bacterial cellulose: a review. Cellulose. 23 (2016) 57–91. https://doi.org/10.1007/s10570-015-0802-0

[19] R. Auta, G. Adamus, M. Kwiecien, I. Radecka, P. Hooley, Production and characterization of bacterial cellulose before and after enzymatic hydrolysis. Afr. J. Biotechnol. 16 (2017) 470-482. https://doi.org/10.5897/AJB2016.15486

[20] G. Pacheco, C.R. Nogueira, A.B. Meneguin, E. Trovatti, M.C.C. Silva, R.T.A. Machado, S.J.L. Ribeiro, E.C.S. Filho, H.S. Baruda, Development and characterization of bacterial cellulose produced by cashew tree residues as alternative carbon source. Ind. Crop. Prod. 107 (2017) 13–19. https://doi.org/10.1016/j.indcrop.2017.05.026

[21] A. Ubeyitogullari, O.N. Ciftci, Formation of nanoporous aerogels from wheat starch.Carbohydr. Polym. 147 (2016) 125-132. https://doi.org/10.1016/j.carbpol.2016.03.086

[22] L. Druel, R. Bard, W. Vorwerg, T. Budtova, Starch aerogels: A member of the family of thermal superinsulating materials.Biomacromolecules. 18 (2017) 4232-4239. https://doi.org/10.1016/j.carbpol.2016.03.086

[23] F. Zhu, Starch based aerogels: Production, properties and applications. Trends Food Sci Tech. 89 (2019) 1–10. https://doi.org/10.1016/j.tifs.2019.05.001

[24] N. Abhari, A. Madadlou, A. Dini, Structure of starch aerogel as affected by crosslinking and feasibility assessment of the aerogel for an anti-fungal volatile release. FoodChem. 221 (2017) 147–152. https://doi.org/10.1016/j.foodchem.2016.10.072

[25] Z. Miao, K. Ding, T. Wu, Z. Liu, B. Han, G. An, S. Miao, G. Yang, Fabrication of 3D-networks of native starch and their application to produce porous inorganic oxide networks through a supercritical route. Micropor. Mesopor. Mat. 111 (2008) 104–109. https://doi.org/10.1016/j.tifs.2019.05.001

[26] H.E. Knidri, R. Belaabed, A. Addaou, A. Laajeb, A. Lahsini, Extraction, chemical modification and characterization of chitin and chitosan. Int. J. Biol. Macromol. 120 (2018) 1181–1189. https://doi.org/10.1016/j.ijbiomac.2018.08.139

[27] N. Hsan, P.K. Dutta, S. Kumar, R. Bera, N. Das, Chitosan grafted graphene oxide aerogel: Synthesis, characterization and carbon dioxide capture study. Int. J. Biol. Macromol 125 (2019) 300–306. https://doi.org/10.1016/j.ijbiomac.2018.12.071

[28] S. Hassan, M. Suzuki, A. A. El-Moneim, Synthesis of MnO_2-chitosan nanocomposite by one-step electrodeposition for electrochemical energy storage

application. J. Power Sources. 246 (2014) 68-73.
https://doi.org/10.1016/j.jpowsour.2013.06.085

[29] S.Takeshita, S. Yoda, Chitosan Aerogels: Transparent, Flexible Thermal Insulators.
Chem. Mater.27(2015) 7569−7572. https://doi.org/10.1021/acs.chemmater.5b03610

[30] C. Li, K. Wu, Y. Su, S.B. Riffat, X. Ni, F. Jiang, Effect of drying temperature on
structural and thermomechanical properties of konjac glucomannan-zein blend films.
Int. J. Biol. Macromol. 138 (2019) 135–143.
https://doi.org/10.1016/j.ijbiomac.2019.07.007

[31] Y. Wang, K. Wu, M. Xiao, S.B. Riffat, Y. Su, F. Jiang, Thermal conductivity,
structure and mechanical properties of konjac glucomannan/starch based aerogel
strengthened by wheat straw. Carbohydr. Polym. 197 (2018) 284–291.
https://doi.org/10.1016/j.carbpol.2018.06.009

[32] C. Xin, J. Chen,H. Liang, J. Wan,J. Li, B. Li, Confirmation and measurement of
hydrophobic interaction in sol-gel system of konjac glucomannan with different degree
of deacetylation. Carbohydr. Polym 174 (2017) 337–342.
https://doi.org/10.1016/j.carbpol.2017.06.088

[33] L. Shao, Y. Cao, Z. Li, W. Hu, S. Li, L. Lu, Dual responsive aerogel made from
thermo pH sensitive graft copolymer alginate-g-P (NIPAM-co-NHMAM) for drug
controlled release. Int. J. Biol. Macromol. 114 (2018) 1338–1344.
https://doi.org/10.1016/j.ijbiomac.2018.03.166

[34] S. Groult, T. Budtova, Thermal conductivity structure correlations in thermal superinsulating pectin aerogels. Carbohydr. Polym. 196 (2018) 73–81.
https://doi.org/10.1016/j.carbpol.2018.05.026

[35] N. Isıklan, S. Tokmak, Microwave based synthesis and spectral characterization of
thermo-sensitive poly (N,N-diethylacrylamide) grafted pectin copolymer. Int. J. Biol.
Macromol. 113 (2018) 669–680. https://doi.org/10.1016/j.ijbiomac.2018.02.155

[36] W. Wang, Y. Fang, X. Ni, K. Wu, Y. Wang, F. Jiang, S.B. Riffat, Fabrication and
characterization of a novel konjac glucomannan-based air filtration aerogels
strengthened by wheat straw and okara. Carbohydr. Polym. 224 (2019) 115-
129.https://doi.org/10.1016/j.carbpol.2019.115129

[37] M. Oschatz, S. Boukhalfa, W. Nickel, J.P. Hofmann, C. Fischer, G. Yushi, S.
Kaskel, Carbide-derived carbon aerogels with tunable pore structure as versatile
electrode material in high power supercapacitors. Carbon. 113 (2017) 283-291.
https://doi.org/10.1016/j.carbon.2016.11.050

[38] C.Y. Zhu, Z.Y. Li, H.Q. Pang, N. Pan, Design and optimization of core/shell structures as highly efficient opacifiers for silica aerogels as high-temperature thermal insulation. Int. J. Therm. Sci. 133 (2018) 206–215. https://doi.org/10.1016/j.ijthermalsci.2018.07.032

[39] T.R. Cuadros, A.A. Erices, J.M. Aguilera, Porous matrix of calcium alginate/gelatin with enhanced properties as scaffold for cell culture, J. Mech. Behav. Biomed. 46 (2015) 331–342. https://doi.org/10.1016/j.jmbbm.2014.08.026

[40] N. Pircher, D. Fischhuber, L. Carbajal, C. Strau, J.M. Nedelec, C. Kasper, T. Rosenau, F.Liebner, Preparation and Reinforcement of Dual-Porous Biocompatible Cellulose Scaffolds for Tissue Engineering. Macromol. Mater. Eng. 300 (2015) 911–924. https://doi.org/10.1002/mame.201500048

[41] Z. Kneza, M. Pantica, D. Cora, Z. Novaka, M.K. Hrncica, Are supercritical fluids solvents for the future?. Chem. Eng. Process. 141 (2019) 107-532. https://doi.org/10.1016/j.cep.2019.107532

[42] G. Brunner, Near critical and supercritical water. Part I. Hydrolytic and hydrothermal processes. J. of Supercritical Fluids. 47 (2009) 373–381. https://doi.org/10.1016/j.supflu.2008.09.002

[43] E.S. Dassoff, Y.O. Li, Mechanisms and effects of ultrasound-assisted supercritical CO_2 extraction. Trends Food Sci Tech. 86 (2019) 492–501. https://doi.org/org/10.1016/j.tifs.2019.03.001

[44] S. Plazzotta, S. Calligaris, L. Manzocco, Structure of oleogels from κ-carrageenan templates as affected by supercritical-CO_2-drying, freeze-drying and lettuce-filler addition. Food Hydrocoll. 96 (2019) 1–10. https://doi.org/10.1016/j.foodhyd.2019.05.008

[45] I. Sahina, E. Uzunlarb, C. Erkeya, Investigation of kinetics of supercritical drying of alginate alcogel particles. J. Supercritic. Fluid. 146 (2019) 78–88. https://doi.org/10.1016/j.supflu.2018.12.019

[46] X. Ni, F. Ke, M. Xiao, K. Wu, Y. Kuanga, H. Corke, F. Jiang, The control of ice crystal growth and effect on porous structure of konjac glucomannan-based aerogels. Int. J. Biol. Macromol. 92 (2016) 1130–1135. https://doi.org/10.1016/j.ijbiomac.2016.08.020

[47] S. Liu, F. Yao, O. Oderinde, Z. Zhang, G. Fu, Green synthesis of oriented xanthan gum–graphene oxide hybrid aerogels for water purification. Carbohydr Polym 174 (2017) 392–399. https://doi.org/10.1016/j.carbpol.2017.06.044

[48] P. Gupta, B. Singh, A.K. Agrawal, P. K. Maji, Low density and high strength nanofibrillated cellulose aerogel for thermal insulation application. Mater. Design. 158 (2018) 224–236. https://doi.org/10.1016/j.matdes.2018.08.031

[49] H. Geng, A facile approach to light weight, high porosity cellulose aerogels. Int. J. Biol. Macromol. 118 (2018) 921–931. https://doi.org/10.1016/j.ijbiomac.2018.06.167

[50] G. Lodhi, Y.S. Kim, J.W. Hwang, S.K. Kim, Y.J. Jeon, J.Y. Je, C.B. Ahn, S.H. Moon, B.T. J.P. Park, Chitooligosaccharide and Its Derivatives: Preparation and Biological Applications. Biomed. Res. Int. 1 (2014) 1-13. https://doi.org/10.1155/2014/654913

[51] C. López-Iglesias, J. Barros, I. Ardao, F.J. Monteiro, C. Alvarez-Lorenzoa, J. L. Gómez-Amozaa, C.A. García-Gonzáleza, Vancomycin-loaded chitosan aerogel particles for chronic wound applications. Carbohydr. Polym. 204 (2019) 223–231. https://doi.org/10.1016/j.carbpol.2018.10.012

[52] H.M. Duong, Z. K. Lim, T. X. Nguyen, B. Gu, M. P. Penefather, N. Phan-Thien, Compressed hybrid cotton aerogels for stopping liquid leakage. Colloid. Surface. A. 537 (2018) 502–507. https://doi.org/10.1016/j.colsurfa.2017.10.067

[53] G. Tkalec, Z. Knez, Z. Novak, Fast production of high-methoxyl pectin aerogels for enhancing the bioavailability of low-soluble drugs. J. of Supercritical Fluids. 106 (2015) 16–22. https://doi.org/10.1016/j.supflu.2015.06.009

[54] G. Horvat, M. Pantic, Z. Knez, Z. Novak, Encapsulation and drug release of poorly water soluble nifedipine from biocarriers. J. Non-Cryst. Solids. 481 (2018) 486–493. https://doi.org/10.1016/j.jnoncrysol.2017.11.037

[55] H. Valo, S. Arola, P. Laaksonen, M. Torkkeli, L. Peltonen, M.B. Linder, R. Serimaa, S. Kuga, J. Hirvonen, T. Laaksonen, Drug release from nanoparticles embedded in four different nanofibrillar cellulose aerogels. Eur. J. Pharm. Sci, 50 (2013) 69–77. https://doi.org/10.1016/j.ejps.2013.02.023

[56] C. Wan, J. Li, Cellulose aerogels functionalized with polypyrrole and silver nanoparticles: In-situ synthesis, characterization and antibacterial activity. Carbohydr. Polym. 146 (2016) 362–367. https://doi.org/10.1016/j.carbpol.2016.03.081

[57] C.A. Bugnone, S. Ronchetti, L. Manna, M. Banchero, An emulsification/internal setting technique for the preparation of coated and uncoated hybrid silica/alginate aerogel beads for controlled drug delivery. J. of Supercritical Fluids. 142 (2018) 1–9. https://doi.org/10.1016/j.supflu.2018.07.007

[58] L. Shao, Y. Cao, Z. Li, W. Hu, S. Li, L. Lu, Dual responsive aerogel made from thermo/pH sensitive graft copolymer alginate-g-P(NIPAM-co-NHMAM) for drug controlled release. Int. J. Biol. Macromol. 114 (2018) 1338–1344. https://doi.org/10.1016/j.ijbiomac.2018.03.166

Aerogels II: Preparation, Properties and Applications   Materials Research Forum LLC
Materials Research Foundations **97** (2021) 23-42   https://doi.org/10.21741/9781644901298-2

Chapter 2

Aerogels for Biomedical Applications

Satyanarayan Pattnaik*, Y. Surendra, J. Venkateshwar Rao, Kalpana Swain

Division of Drug Delivery Systems, Talla Padmavathi College of Pharmacy, Warangal, India

Abstract

The researchers across the world are actively engaged in strategic development of new porous aerogel materials for possible application of these extraordinary materials in the biomedical field. Due to their excellent porosity and established biocompatibility, aerogels are now emerging as viable solutions for drug delivery and other biomedical applications. This chapter aims to cover the diverse aerogel materials used across the globe for different biomedical applications including drug delivery, implantable devices, regenerative medicine encompassing tissue engineering and bone regeneration, and biosensing.

Keywords

Aerogel, Biomedicine, Regenerative Medicine, Drug Delivery, Tissue Engineering, Biosensing

Contents

1.   Introduction

Nanotechnology has evolved steadily offering a diversity of solutions for optimal drug delivery leading to improved therapeutic outcomes [1-5]. Over the last few decades, sincere efforts have been made to develop several porous materials for possible exploitations in the field of biomedicine [6-8]. The quest for the porous materials has triggered researchers worldwide for the strategic development of aerogels with an intention of possible deployment of these extraordinary groups of materials (aerogels) in biomedicine [9-11]. Technically, aerogels are solid foams with very little solid and up to 99.8 % air [11-13]. This is why these materials are ultra-light and the unique composition gives them a ghostly appearance, hence it is often referred to as "frozen smoke". Owing to extremely high porosity and proven biocompatibility, aerogels are now offering viable solutions for drug delivery and other biomedical applications.

Aerogels are usually fabricated following sol-gel polymerization, wherein air replaces the solvent system adopting an efficient evaporation technique. Adopting this method, a three-dimensional ultra-lightweight porous nanostructure is prepared having nearly similar dimensions as that of the original dispersion. Aerogel types may be broadly classified based on the precursor types as silica-based aerogels, organic-inorganic silica aerogels, organic aerogels and bio-aerogels.

The use of aerogels in biomedicine is a relatively naïve concept. Various physicochemical properties of aerogels like large porosity, tunable pore diameter, huge surface area, biodegradability, and established biocompatibility has pooled interest of researchers for diverse biomedical applications like biomedical implantable devices [14,15], bone regeneration and tissue engineering [16-20], biosensing and diagnosis [21-24], and drug delivery [25-29].

This chapter aims at providing a gist of various types of aerogels deployed by researchers worldwide for biomedical applications.

2.   Aerogels in biomedicine

Based on their composition aerogels are usually categorized as inorganic, organic and hybrid aerogels [30]. Alkoxides and metal oxides are usually used as starting materials for fabrication of inorganic aerogels which forms the most diverse category of aerogels. Amongst all these, silica-based aerogels are the most widely investigated products. Synthesis of organic aerogels date back to the year 1987. Resorcinol-formaldehyde aerogels are the most studied materials under the organic aerogel category and are stiffer and stronger than silica aerogels. However, toxicity of the chemical used for its preparation limits its applications in biomedical industry.

Aerogels II: Preparation, Properties and Applications   Materials Research Forum LLC
Materials Research Foundations **97** (2021) 23-42   https://doi.org/10.21741/9781644901298-2

Organic–inorganic hybrid aerogels are relatively newer in the industry. Synthesis of these aerogel materials presents interesting possibilities to blend the characteristics of both the organic and inorganic materials yielding desired hybrid. The various biomedical applications of these aerogel materials are dealt with here.

2.1   Aerogels in drug delivery

Many aerogels are biodegradable and biocompatible and hence are suitable carriers for delivery of drugs. Optimal drug delivery strategies usually aim at delivering the cargo at the desired target site at an appropriate concentration [2-4, 31-34]. While delivering drugs enough care must be taken to minimize any possible adverse effects of the drug [35]. The drug release mechanism is often an important factor that warrants attention of the researchers while developing drug delivery systems [36-38]. The release mechanism can be optimally fine-tuned by controlling the process parameters during fabrication of delivery systems or by altering the formulation variables [39]. Like other porous carriers for drug delivery, aerogel materials present a huge internal pore volume for possible drug loading and delivery.

Obaidat and co-workers [26] used the emulsion-gelation technique for the fabrication of carrageenan aerogel microparticles for the delivery of ibuprofen. The basic objective of the delivery system was to improve the dissolution velocity of the poorly soluble drug candidate. The authors were able to load ibuprofen in its amorphous state in the porous structure of carrageenan aerogel microparticles. Subsequent studies revealed a significant improvement of dissolution indicating the carrageenan aerogel microparticles as potential carriers for poorly soluble drugs. Polysaccharides (starch, pectin, and alginate) aerogel microspheres were used as a successful carrier for delivery of another poorly soluble drug candidate, ketoprofen [40]. Drug release patterns from the pectin and alginate aerogels followed the Gallagher-Corrigan release model whereas drug release from starch aerogels followed first-order release kinetics.

Intriguingly, many polysaccharides have muco-adhesive properties and hence can be used for mucosal drug delivery. Alginate-based hybrid aerogel microparticles were deployed as carriers for the mucosal administration of ketoprofen and quercetin [41]. The hybrid aerogels were prepared by co-gelation of low methoxyl pectin and κ-carrageenan with alginate and subsequent drying with supercritical CO_2. The drug loading was carried out using supercritical CO_2 adsorption (for ketoprofen) or supercritical antisolvent precipitation (for quercetin). κ-carrageenan hybrid aerogel was found superior in terms of rate of drug release.

Chitosan and carboxymethyl cellulose are pH-sensitive polymers and hence they are often used for stimuli (pH) responsive drug delivery. Controlled delivery of anticancer agent 5-fluorouracil was materialized using calcium ion crosslinked hybrid chitosan aerogels (CS/CMC/Ca^{2+}/GO) [42]. Analysis of drug release pattern revealed release of 5-fluorouracilfollowing Fick's diffusion model.

In another attempt to investigate the aerogel platform to deliver poorly soluble drug candidates, Veres et al. [43] fabricated a series of fourteen hybrid aerogel products containing silica and gelatin with proper surface. For the studied poorly soluble drugs, modified release was evidenced by modulating the composition of aerogel matrix and surface treatments.

Shao et al. [27] successfully fabricated a dual responsive (temperature and pH) alginate-g-P(NIPAM-co-NHMAM) aerogel for controlled drug delivery with drug loading efficiency up to 13.24%. There are many reports for sustained delivery of highly soluble drugs using aerogels. Aerogels fabricated from polyethylenimine-grafted cellulose nanofibrils (CNFs-PEI) were developed with improved cargo loading abilities (287.3 mg/g) for sustained delivery of a soluble drug candidate sodium salicylate [44]. The drug sorption process in the hybrid aerogels was well described by Langmuir equation. Drug release from the delivery system exhibited a sustained release behavior dependent on pH and temperature. In another interesting work, ibuprofen was delivered using Fe(III)-crosslinked alginate aerogels [45]. Faster ibuprofen release was reported in a basic dissolution media (pH-7.4) than in acidic media (pH-2). Incorporation of ascorbic acid in the aerogels significantly improved the ibuprofen release in both the types of dissolution media which is probably due to reduction of crosslinker Fe (III) to Fe(II) leading to poor interaction with alginate. Subsequently, the chains get hydrated causing the matrix to erode and dissolve. Recently, alginate-based aerogel particles were developed and evaluated for possible delivery of few poorly soluble drug candidates [46]. X-ray diffractograms confirmed stable amorphous state of loaded drugs inside the aerogels. Dissolution studies indicated an improvement in the dissolution velocity of all the three studied drugs (Fig 1 [46]). In an attempt to improve the oral bioavailability of paclitaxel, nanoporous silica aerogels were developed [47]. The researchers claimed an improved bioavailability, reduced side effects of the drug and inhibited tumor growth.

Dysprosia aerogels, for drug cargo delivery, were synthesized and subsequently reinforced by a thin polyurea coating applied over their skeletal structure [48]. The scattered mesopores of dysprosia aerogels were loaded with up to 30% v/v with drug (paracetamol/ indomethacin/insulin). These aerogel materials owing to their high drug loading capabilities and sustained release properties augmented with relatively lower toxicity, significant magnetic effect, and the probability for neutron activation are ideal carriers for temporal drug delivery.

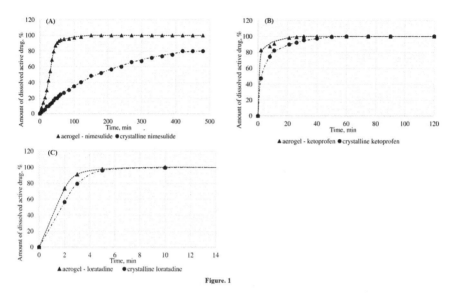

Figure. 1

Fig. 1 In vitro dissolution profile of (A)nimesulide loaded alginate-based aerogel, (B) ketoprofen loaded alginate-based aerogel, and (C) loratadine loaded alginate-based aerogel. Adapted from Lovskaya et al. [46].

Water-insoluble aerogels were synthesized from laccase-oxidized galactomannans offenugreek and assessed for delivery of lysozyme (antimicrobial active) [49]. The activity of lysozyme was not adversely affected during its processing evidenced by its hydrolytic glycosidase activity on lyophilized Micrococcus lysodeikticus cells wall peptidoglycans. This material may present as a carrier for drug delivery and may find future potential applications in nutraceutical delivery. Mohammadian et al. [50], in an attempt to deliver poorly soluble ketoprofen using silica aerogels, reported significant improvement ($p < 0.05$) in the dissolution of loaded ketoprofen. The density of the prepared aerogels seemed to influence the release of trapped drug molecules with highest drug release from the low density aerogel samples. The mechanism of ketoprofen release from the aerogels followed the first-order model. Silica hybrid aerogel materials with differing gelatin content (from 4-24%w/w) were evaluated as carriers for the delivery of a few poorly soluble drug candidates [51]. Supercritical fluid assisted drug loading induced amorphization of drugs inside the porous structure of the aerogels. The concentration of gelatin controlled the release of active ingredients from the drug delivery device with lower gelatin content exhibited rapid release of drugs.

Controlled delivery of methotrexate was possible through the fabrication of hybrid silica aerogel device which exhibited very high cytotoxicity against in vitro tumor cell lines [52]. Successful attempts were made for the controlled delivery of amoxicillin via loading into cellulose aerogels [53]. Drug dissolution studies indicated the controlled delivery of amoxicillin and antibacterial assay exhibited significant antibacterial activity. Another antimicrobial agent, vancomycin HCl was also successfully delivered through chitosan aerogel microparticles [54]. The developed aerogel microparticles exhibited superior texture, capable of absorbing large volume of wound exudates, and sustained delivery of the antimicrobial agent.

2.2   Aerogels in biomedical implantable devices

Biomedical implants are biocompatible artificial medical devices fabricated to surrogate a missing/underactive/damaged biological structure. Ideal implants should not cause any undesired reactions from neighboring or distant tissues. In certain cases, the implant and surrounding tissue interactions lead to complications similar to any other invasive surgical procedure including infection, inflammation, pain, immunological foreign body response, and implant rejection. The degree of severity of complications varies with the type of implant and site of application in the body. Hence, utmost care is needed while developing biomedical implants to minimize such complications. Recently, the revolutionary developments in nanotechnology have offered many biomaterials for possible application as implantable devices. Aerogel materials are also widely investigated for such application.

Sabri et al. [55], assessed polyurea crosslinked silica aerogels (PCSA) for possible neuronal scaffold application. Adherence and survival of nerves on PCSA coated with various materials were investigated. Untreated polyurea crosslinked silica aerogel surfaces did not favor survival of the nerves, however, coating of laminin1 on the aerogel surface improved adhesion of the cells and supported growth and differentiation of neurons (Fig.2) [55]. Reports from the same laboratory also assessed the biocompatibility of PCSA implants in the rat model [56]. The aerogel implants were well condoned as *s.c/i.m* implants in the Sprague-Dawley rat up to 20 months of incubation period. There was no significant evidence of tissue toxicity confirming that silica-based aerogel could be exploited as biomaterials (Fig. 3) [56]. These PCSA materials were used for repair of severed peripheral nerves in Sprague-Dawley rat allowing a novel option for nerve tissue engineering [57]. The extension of neurites by PC12 cells plated on matrigel-coated and collagen-coated mesoporous polyurea crosslinked silica aerogel surfaces were investigated [58]. The results of the investigations have indicated the possible application of the class of aerogels for the development of neural implants.

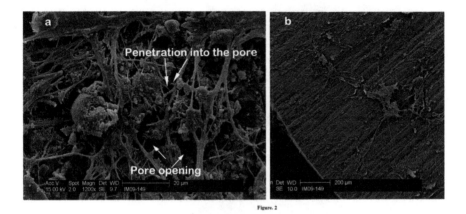

Figure. 2

Fig.2 SEM image of neurons connecting on laminin-coated PCSA.
(a) A dense array of processes has extended across the area of PCSA covered by laminin.Several layers of nerve cells with crisscrossing neuritis can be observed. The pores of the PCSA can be clearly seen and some processes may have extended into the pores as indicated by the arrows. (b) SEM image of the edge of a PCSA disc+Laminin+neurons showing the grooves created by the saw blade. It is hypothesized that the non-homogeneity of the surface provides better anchoring and attachment opportunities for the cell body and the processes to be extended. Adapted from Sabri et al. [55].

The biocompatibility of polyurea-nano-encapsulated surfactant-templated aerogels was investigated for possible biomedical applications [59].The researchers have assessed the compatibility of these materials with components of the vascular system. The researchers found any platelet activation incidences upon contact. The aerogels induced no changes in plasma C3a levels indicating the absence of any inflammation. Moreover, no structural changes or protein deposition on the aerogel samples were confirmed. The reports suggested possibilities of exploitation of these aerogels for applications in cardiovascular implantable devices. Yet, in another study to assess the biocompatibility of aerogel materials, the biocompatibility of Desmodur N3300-derived polyurea aerogels was studied [60]. The results were promising for N3300 aerogels indicating compatibility with endothelial cells with no alteration in blood cells, no inflammatory responses and no absorption of proteins.

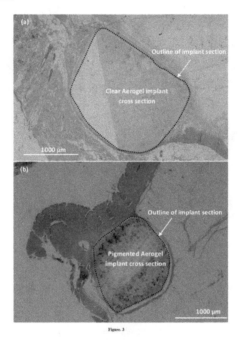

Figure. 3

Fig.3 Histological evaluation of the long-term effect of subcutaneous PCSA implants on nearby tissue.Histology of (a) clear and (b) pigmented aerogel implant extracted after seventeen months of subcutaneous incubation in Sprague-Dawley rats and stained with H&E. A mild fibrosis is observed but no inflammation. The dotted line outlines the boundary between PCSA and nearby tissue. Adapted from Sabri et al. [56].

2.3 Aerogels in tissue engineering and bone regeneration

Tissue engineering has evolved as a discipline from the extensive research carried out in the field of biomaterials. It involves conjugation of scaffolds, cells, and other biologically active molecules to develop functional tissue to augment restoration, maintenance, and improvement of damaged tissue. Regeneration of bone is a very complex physiological process of bone formation that usually happens during normal healing of bone fractures.

The fundamental science of tissue engineering and bone regeneration are often overlapping [2].

Bilayer aerogel scaffold containing cellulose nanofiber/poly (vinyl) alcohol (CNF/PVA) were evaluated for potential skin tissue engineering applications [61]. The biocompatibility of the scaffolds was confirmed by MTT assay indicating the suitability of the material for skin repair applications.

Silk fibroin (SF) has been widely exploited in biomedicine owing to its biodegradability and biocompatibility properties. Mallepally et al. [62] used CO_2 assisted acidification for the synthesis of silk fibroin hydrogels followed by conversion to SF aerogels. The physicomechanical properties were influenced by the concentration of aqueous fibroin. The cytocompatibility of the SF aerogel scaffold was exhibited in cell culture studies with human fore-skin fibroblast cell. Three-dimensional silica-silk fibroin hybrid aerogel scaffolds were synthesized and assessed for bone regeneration applications [63].Intriguingly, these hybrid aerogel could be ideal for bone tissue regeneration. Radiologic and tomographic studies exhibited an interesting osteointegration process.

A highly porous hybrid aerogel consists of graphene oxide (GO)and type I collagen (COL) using sol-gel process (concentrations of GO: 0, 0.05, 0.1, and 0.2% w/v) was fabricated for bone regeneration applications [16]. Graphene oxide concentration in the aerogel influenced the compressive modulus. The hybrid aerogel scaffolds showed promising biomineralization effects in rat cranial defect models.

In another attempt to develop hydrogel scaffold, polyethylene glycol diacrylate (PEGDA) and cellulose nono fibril (CNF)were mixed and visible light photoinitiator was added to form a bio-resin. Further, stereolithography and freeze-drying are put forward to fabricate the aerogel-wet hydrogel scaffold [64]. Three-dimensional (3D) nano-porous scaffolds fabricated fromnatural polymers have recently drawn sizable attention in tissue engineering. The researchers have also reported that the CNFs/PEGDA mixtures with different CNFs contents were all transparent, homogeneous and with obvious shear-thinning property [65]. The NIH 3T3 cells tightly adhered to the CNFs/PEGDA materials and spread on the scaffolds with good differentiation and viability.

Liu et al. [66] developed a nano-cellulose-based aerogels for tissue engineering applications. The charge density, swelling media conditions, and processing parameters were found to influence the degree of swelling and porosity of the nanocelluloses. In vitro cell-based assays revealed very promising results. Since the surface of the polybenzoxazine (PBO)-based aerogels are in close resemblance to the extracellular matrix of bone, their biocompatibility was assessed for possible use as scaffold for bone tissue engineering [67]. Polybenzoxazine-derived carbon aerogels exhibited desired biocompatibility towards osteoblasts and were found to be highly dense, porous and superior mechanical properties. The repair capacity of the articular cartilage tissues is very limited and hence there is a strong need to develop tissue engineered scaffolds for treatment of orthopedic trauma and other joint diseases. Intriguingly, a novel hybrid scaffold composed of methacrylated chondroitin sulfate (CSMA), polyethylene glycol methyl ether-ε-caprolactone-acryloyl chloride and graphene oxide (GO) was fabricated and evaluated for potential cartilage tissue engineering applications [68]. The fabricated scaffold was physico-mechanically identical to the cartilage matrix. The cartilage cells had shown suitable growth potential on the scaffold. The biocompatibility assessment data of the hybrid scaffolds using 3T3 cells were very promising (Fig 4 A) [68]. In vivo biocompatibility and biodegradability was assessed by implanting the scaffold subcutaneously in a rat model (Fig 4 B-E) [68].

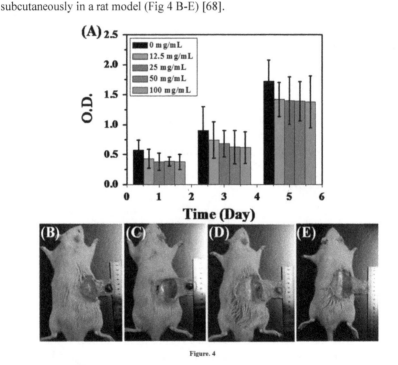

Figure. 4

Fig.4 (A) Cell viability of 3T3 cells in the leachates of scaffold (0, 12.5, 25, 50, 100 mg/mL) at 1, 3 and 5 days. The photographs of subcutaneous implanted CSMA/PECA/GO scaffold on the back of mice for (B) 1, (C) 2, (D) 4 and (E) 8 weeks. Adapted from Liao et al. [68].

2.4   Aerogels in biosensing

Biosensing refers to easy and rapid detection and measurement of target molecules in body fluids using compact devices which acquire a measurable signal obtained by a molecular detection device and a transducer. Developments in the field of nanobiotechnology and microelectronics have contributed a lot for designing of biosensors. Aerogel materials have been widely exploited for biosensing applications. Method to design a highly sensitive quantum dots aerogel-based biosensor for organophosphorous sensing was recently reported [69]. The QDs-AChE aerogel based microfluidic arrays sensor provided good sensitivity for rapid detection of OPs with a detection limit of 0.38 pM. Jeong et al. [70] reported a graphene aerogel-based glucose biosensor reinforced with MoS_2. The three-dimensional aerogel offered platform for enzyme immobilization with high sensitivity and fast response.

Detection of cancerous cell is a growingly important avenue in biosensing research for early diagnosis of cancers to prevent metastasis and better and early treatment reducing mortality. Graphene aerogel microspheres were reported with suitable functionalization with octadecylamine for tumor cell detection [71]. Palladium@gold nanoalloys/nitrogen and sulphur-functionalized multiple graphene aerogel composite (Pd@Au/N,S-MGA) were fabricated for biosensing of dopamine [72]. Researchers have also reported hybrid carbon aerogel materials for biosensing of H_2O_2 [73]. The fabricated sensors exhibited high sensitivity and a low-detection limit. Interesting hybrid graphene aerogel-based materials were reported for glucose-biosensing applications [74]. The gold nanoparticle doped hybrid graphene aerogels were reportedly very sensitive for glucose oxidase immobilization.

Three-dimensional $Cu@Cu_2O$ aerogels were developed for voltammetric sensing of glucose [75]. The sensors were reportedly highly efficient with a detection limit of 54 μM (S/N = 3). Polydopamine functionalized metal-doped aerogels were reported for immobilization of acetylcholinesterase in biosensing of organophosphorous pesticides [76].

3.   Future prospects and conclusions

The recent developments in aerogel research are expected to have sound technoscientific and socioeconomic impact through development of value-added products, biologically safe and environment friendly technologies. Usages of bioresources as raw materials and green solvents for processing of the aerogels have significantly reduced the risks for regulatory compliance. However, there still exist an acute need for devising scale-up

Materials Research Forum LLC
https://doi.org/10.21741/9781644901298-2

strategies developing products from laboratory scale to commercial products taking into considerations of economic and safety aspects.

Fabrication of the new generation aerogels with desired properties is expected to overcome the many of the physicochemical and physicomechanical hurdles of present biomedical technology. The tailorable chemical composition and highly porous network together with the excellent mechanical properties of some aerogel materials have satisfied the tissue engineering application requirement that is not easily achievable by other porous biomaterials. A plethora of attempts are still in progress to exploit the aerogel composites for fabrication of cardiovascular implants as well as for nerve repair implant applications. Research has established that aerogel based materials are very promising as drug delivery vehicles due to their high drug loading capabilities, ability for controlled/sustained and tunable drug release, capability to increase the oral bioavailability of poorly soluble drugs, and to improve drug stability. Taking into account the increasing importance of aerogels in a variety of application areas, their use in drug delivery and other biomedical avenues can be expected to grow in the coming years.

References

[1] S.S. Hota, S.Pattnaik, S.Mallick,Formulation and evaluation of multidose propofol nanoemulsion using statistically designed experiments, Acta ChimicaSlovenica. 67 (2020) 179-188. https://doi.org/10.17344/acsi.2019.5311

[2] S. Pattnaik, K. Swain, Z. Lin, Graphene and graphene-based nanocomposites: biomedical applications and biosafety, J. Mater. Chem. B. 4(2016) 7813-7831. https://doi.org/10.1039/C6TB02086K

[3] S. Pattnaik, K.Swain, J.V.Rao, T.Varun, S.K.Subudhi, Aceclofenac nanocrystals for improved dissolution: influence of polymeric stabilizers, RSC Adv. 5(2015) 91960-91965. https://doi.org/10.1039/C5RA20411A

[4] S. Pattnaik, K.Swain, P.Manaswini, E.Divyavani, J.V.Rao, T. Varun, S.K.Subudhi. Fabrication of aceclofenac nanocrystals for improved dissolution: Process optimization and physicochemical characterization, J Drug Deliv Sci Technol. 29(2015) 199-209. https://doi.org/10.1016/j.jddst.2015.07.021

[5] K.Pathak, S.Pattnaik, K.Swain, Application of nanoemulsions in drug delivery. In: S.M. Jafari, D. J. McClements (Eds.), Nanoemulsions: Formulation, Applications and Characterization, Academic Press (Elsevier), Amsterdam,2018, pp. 415-433. https://doi.org/10.1016/B978-0-12-811838-2.00013-8

Materials Research Forum LLC
https://doi.org/10.21741/9781644901298-2

[6] Y.C. Yadav, S. Pattnaik, K. Swain, Curcumin loaded mesoporous silica nanoparticles: Assessment of bioavailability and cardioprotective effect, Drug Dev Ind Pharm. 45(2019) 1889-1895. https://doi.org/10.1080/03639045.2019.1672717

[7] S. Pattnaik, K. Pathak, Mesoporous silica molecular sieve based nanocarriers: Transpiring drug dissolution research. Curr. Pharm. Des.23 (2017) 467-480. https://doi.org/10.2174/1381612822666161026162005

[8] S. Pattnaik, K. Swain, Mesoporous nanomaterials as carriers in drug delivery. In: Inamuddin,A.M. Asiri, A. Mohammad (Eds.), Applications of Nanocomposite Materials in Drug Delivery,Woodhead Publishing (Elsevier),Cambridge, 2018, pp. 589-604. https://doi.org/10.1016/B978-0-12-813741-3.00025-X

[9] Y. Ma, Y. Yue, H. Zhang, F. Cheng, W. Zhao, J. Rao, S. Luo, J. Wang, X. Jiang, Z. Liu, N. Liu,Y. Gao, 3D synergistical MXene/Reduced graphene oxide aerogel for a piezoresistive sensor.ACS Nano. 12 (2018) 3209-3216. https://doi.org/10.1021/acsnano.7b06909

[10]   L. Wang, R.J. Mu, L. Lin, X. Chen, S. Lin, Q. Ye, J. Pang,Bioinspired aerogel based on konjac glucomannan and functionalized carbon nanotube for controlled drug release.Int. J. Biol. Macromol. 133(2019) 693-701. https://doi.org/10.1016/j.ijbiomac.2019.04.148

[11]   C.A. García-González, T. Budtova, L. Durães, C. Erkey, P. Del Gaudio, P. Gurikov,M. Koebel, F. Liebner, M. Neagu, I. Smirnova, An opinion paper on aerogels for biomedical and environmental applications, Molecules. 24 (2019) pii: E1815. https://doi.org/10.3390/molecules2409181510.3390/molecules24091815.

[12]   S.S. Kistler,Coherent expanded aerogels and jellies, Nature. 127 (1931) 741-741. https://doi.org/10.1038/127741a0

[13]   S.S. Kistler, Coherent expanded-aerogels, J. Phys. Chem. 36(1932) 52-64. https://doi.org/10.1021/j150331a003

[14]   F. Sabri, M.E. Sebelik, R. Meacham, J. D. Jr Boughter, M.J. Challis, N. Leventis, In vivo ultrasonic detection of polyurea crosslinked silica aerogel implants. PLoS One. 8(2013) e66348. https://doi.org/10.1371/journal.pone.0066348

[15]   H.R. Stanley, M.B. Hall, A.E. Clark, C.J. King 3rd, L.L. Hench, J.J.Berte, Using 45S5 Bioglass Cones as endosseous ridge maintenance implants to prevent alveolar ridge resorption: A 5-year evaluation, Int. J. Oral. Maxillofac. Implants. 12(1997) 95-105.

[16]   S. Liu, C. Zhou, S. Mou, J. Li, M. Zhou, Y. Zeng, C. Luo, J. Sun J, Z. Wang, W. Xu, Biocompatible graphene oxide-collagen composite aerogel for enhanced stiffness and in situ bone regeneration. Mater Sci Eng C Mater Biol Appl. 105 (2019) 110137. https://doi.org/10.1016/j.msec.2019.110137

Materials Research Forum LLC
https://doi.org/10.21741/9781644901298-2

[17]   H. Maleki, M.A. Shahbazi, S. Montes, S.H. Hosseini, M.R. Eskandari, S. Zaunschirm, T. Verwanger, S. Mathur, B. Milow, B. Krammer, N. Hüsing, Mechanically strong silica-silk fibroin bioaerogel: A hybrid scaffold with ordered honeycomb micromorphology and multiscale porosity for bone regeneration, ACS Appl. Mater. Interfaces. 11 (2019) 17256-17269. https://doi.org/10.1021/acsami.9b04283

[18]   S. Dong, Y.N. Zhang, J. Wan, R. Cui, X. Yu, G. Zhao, K. Lin, A novel multifunctional carbon aerogel-coated platform for osteosarcoma therapy and enhanced bone regeneration, J. Mater. Chem. B. 22 (2020)368-379. https://doi.org/10.1039/C9TB02383F

[19]   R. Ghafari, M. Jonoobi, L.M. Amirabad, K. Oksman, A.R. Taheri, Fabrication and characterization of novel bilayer scaffold from nanocellulose based aerogel for skin tissue engineering applications, Int. J. Biol. Macromol. 136 (2019) 796-803. https://doi.org/10.1016/j.ijbiomac.2019.06.104

[20]   D.A. Osorio, B.E.J. Lee, J.M. Kwiecien, X. Wang, I. Shahid, A.L. Hurley, E.D. Cranston, K. Grandfield, Cross-linked cellulose nanocrystal aerogels as viable bone tissue scaffolds, Acta Biomater. 87 (2019) 152-165. https://doi.org/10.1016/j.actbio.2019.01.049

[21]   I. Ali, L. Chen, Y. Huang, L. Song, X. Lu, B. Liu, L. Zhang, J. Zhang, L. Hou, T. Chen, Humidity-responsive gold aerogel for real-time monitoring of human breath, Langmuir.34 (2018) 4908-4913. https://doi.org/10.1021/acs.langmuir.8b00472

[22]   L. Dong, W. Wang, J. Chen, X. Ding, B. Fang, X. Miao, Y. Liu, F. Yu, H. Xin, X. Wang, Silver nanowire net knitted anisotropic aerogel as an ultralight and sensitive physiological activity monitor, Biomater. Sci. 6(2018) 2312-2315. https://doi.org/10.1039/C8BM00651B

[23]   X. Niu, X. Li, W. Chen, X. Li, W. Weng, C. Yin, R. Dong, W. Sun, G. Li.Three-dimensional reduced graphene oxide aerogel modified electrode for the sensitivequercetin sensing and its application, Mater. Sci. Eng. C. Mater. Biol. Appl. 89(2018)230-236. https://doi.org/10.1016/j.msec.2018.04.015

[24]   L. Ruiyi, C. Fangchao, Z. Haiyan, S. Xiulan, L. Zaijun.Electrochemical sensor for detection of cancer cell based on folic acid and octadecylamine-functionalized grapheneaerogel microspheres, Biosens. Bioelectron. 119 (2018) 156-162. https://doi.org/10.1016/j.bios.2018.07.060

[25]   G. Vasvári, J. Kalmár, P. Veres, M. Vecsernyés, I. Bácskay, P. Fehér, Z. Ujhelyi, A. Haimhoffer, A. Rusznyák, F. Fenyvesi, J. Váradi, Matrix systems for oral drug delivery: Formulations and drug release, Drug Discov. Today Technol. 27(2018) 71-80. https://doi.org/10.1016/j.ddtec.2018.06.009

[26]   R.M. Obaidat, M. Alnaief, H. Mashaqbeh, Investigation of carrageenan aerogel microparticles as a potential drug carrier.AAPS Pharm.Sci.Tech. 19 (2018) 2226-2236. https://doi.org/10.1208/s12249-018-1021-4

[27]   L. Shao, Y. Cao, Z. Li, W. Hu, S. Li, L. Lu, Dual responsive aerogel made from thermo/pH sensitive graft copolymer alginate-g-P(NIPAM-co-NHMAM) for drug controlled release, Int. J. Biol. Macromol. 114(2018) 1338-1344. https://doi.org/10.1016/j.ijbiomac.2018.03.166

[28]   P. Veres, M. Kéri, I. Bányai, I. Lázár, I. Fábián, C. Domingo, J. Kalmár, Mechanism of drug release from silica-gelatin aerogel-relationship between matrix structure and release kinetics, Colloids Surf. B Biointerfaces. 152 (2017) 229-237. https://doi.org/10.1016/j.colsurfb.2017.01.019

[29]   X. Wang, J. Wang, S. Feng, Z. Zhang, C. Wu, X. Zhang, F. Kang, Nano-porous silica aerogels as promising biomaterials for oral drug delivery of paclitaxel, J. Biomed. Nanotechnol. 15 (2019) 1532-1545. https://doi.org/10.1166/jbn.2019.2763

[30]   Z. Ulker, C. Erkey, An emerging platform for drug delivery: aerogel based systems,J. Control Rel. 177 (2014) 51-63. https://doi.org/10.1016/j.jconrel.2013.12.033

[31]   S. Pattnaik, K. Swain, J.V. Rao, T. Varun, S. Mallick. Temperature influencing permeation pattern of alfuzosin: An investigation using DoE, Medicina. 51(2015) 253-261. https://doi.org/10.1016/j.medici.2015.07.002

[32]   S. Pattnaik, K. Swain, J.V. Rao, T. Varun, K.B. Prusty, S.K. Subudhi, Polymer co-processing of ibuprofen through compaction for improved oral absorption, RSC Adv. 5(2015)74720-74725. https://doi.org/10.1039/C5RA13038G

[33]   A.K. Mahapatra, P.N. Murthy, R.K. Patra, S. Pattnaik, Solubility enhancement of modafinil by complexation with β-cyclodextrin and hydroxypropyl β-cyclodextrin: aresponse surface modeling approach, Drug Delivery Letters. 3(2013) 210-219. https://doi.org/10.2174/22103031113039990005

[34]   S. Pattnaik, K. Swain, S. Mallick, Z. Lin, Effect of casting solvent on crystallinity of ondansetron in transdermal films, Int. J. Pharm. 406(2011) 106-110. https://doi.org/10.1016/j.ijpharm.2011.01.009

[35]   K. Pathak, S. Pattnaik,A. Porwal,Regulatory concerns for nanomaterials in sunscreen formulations, Applied Clinical Research, Clinical Trials and Regulatory Affairs. 5(2018) 99-111. https://doi.org/10.2174/2213476X05666180601103853

[36]   K. Swain, S. Pattnaik, N. Yeasmin, S. Mallick, Preclinical evaluation of drug in adhesive type ondansetron loaded transdermal therapeutic systems, Eur. J. Drug Metab. Pharmacokinet. 36(2011) 237-241. https://doi.org/10.1007/s13318-011-0053-x

[37]   S. Pattnaik, K. Swain, A. Bindhani, S. Mallick, Influence of chemical permeation enhancers on transdermal permeation of alfuzosin: An investigation using response surface modeling, Drug. Dev. Ind. Pharm. 37(2011) 465-474. https://doi.org/10.3109/03639045.2010.522192

[38]   K. Swain, S. Pattnaik, S. Mallick, K.A. Chowdary, Influence of hydroxypropyl methylcellulose on drug release pattern of a gastroretentive floating drug delivery system using a 3^2 full factorial design, Pharm. Dev. Technol. 14(2009)193-198. https://doi.org/10.1080/10837450802498902

[39]   K. Swain, S. Pattnaik, S.C. Sahu, K.K. Patnaik, S. Mallick, Drug in adhesive type transdermal matrix systems of ondansetron hydrochloride: optimization of permeation pattern using response surface methodology, J. Drug. Target. 18(2010) 106-114. https://doi.org/10.3109/10611860903225727

[40]   C.A. García-González, M. Jin, J. Gerth, C. Alvarez-Lorenzo, I. Smirnova, Polysaccharide-based aerogel microspheres for oral drug delivery, CarbohydrPolym. 117(2015) 797-806. https://doi.org/10.1016/j.carbpol.2014.10.045

[41]   V.S. Gonçalves, P. Gurikov, J. Poejo, A.A. Matias, S. Heinrich, C.M. Duarte, I. Smirnova, Alginate-based hybrid aerogel microparticles for mucosal drug delivery, Eur. J. Pharm. Biopharm. 107(2016) 160-70. https://doi.org/10.1016/j.ejpb.2016.07.003

[42]   R. Wang, D. Shou, O. Lv, Y. Kong, L. Deng, J. Shen, pH-Controlled drug delivery with hybrid aerogel of chitosan, carboxymethyl cellulose and graphene oxide as the carrier, Int. J. Biol. Macromol. 103(2017) 248-253. https://doi.org/10.1016/j.ijbiomac.2017.05.064

[43]   P. Veres, A.M. López-Periago, I. Lázár, J. Saurina, C. Domingo, Hybrid aerogel preparations as drug delivery matrices for low water-solubility drugs, Int. J. Pharm. 496(2015) 360-370. https://doi.org/10.1016/j.ijpharm.2015.10.045

[44]   J. Zhao, C. Lu, X. He, X. Zhang, W. Zhang, X. Zhang, Polyethylenimine-grafted cellulose nanofibril aerogels as versatile vehicles for drug delivery, ACS Appl. Mater. Interfaces. 7(2015) 2607-2615. https://doi.org/10.1021/am507601m

[45]   P. Veres, D. Sebők, I. Dékány, P. Gurikov, I. Smirnova, I. Fábián, J. Kalmár, A redox strategy to tailor the release properties of Fe(III)-alginate aerogels for oral drug delivery, CarbohydrPolym. 188(2018)159-167. https://doi.org/10.1016/j.carbpol.2018.01.098

[46]   D. Lovskaya, N. Menshutina, Alginate-based aerogel particles as drug delivery systems: investigation of the supercritical adsorption and in vitro evaluations,Materials (Basel). 13(2020) pii: E329. https://doi.org/10.3390/ma13020329

[47]   X. Wang, J. Wang, S. Feng, Z. Zhang, C.Wu, X. Zhang, F. Kang, Nano-porous silica aerogels as promising biomaterials for oral drug delivery of paclitaxel, J. Biomed. Nanotechnol. 15 (2019) 1532-1545. https://doi.org/10.1166/jbn.2019.2763

[48]   A. Bang, A.G. Sadekar, C. Buback, B. Curtin, S. Acar, D. Kolasinac, W. Yin, D.A. Rubenstein, H. Lu, N. Leventis, C. Sotiriou-Leventis, Evaluation of dysprosia aerogels as drug delivery systems: a comparative study with random and ordered mesoporous silicas, ACS Appl. Mater. Interfaces. 6(2014) 4891-4902. https://doi.org/10.1021/am4059217

[49]   B. Rossi, P. Campia, L. Merlini, M. Brasca, N. Pastori, S. Farris, L. Melone,C. Punta, Y.M. Galante. An aerogel obtained from chemo-enzymatically oxidized fenugreek galactomannans as a versatile delivery system, CarbohydrPolym. 144(2016) 353-361. https://doi.org/10.1016/j.carbpol.2016.02.007

[50]   M. Mohammadian, T.S.J.Kashi, M. Erfan, F.P. Soorbaghi, In-vitro study of ketoprofen release from synthesized silica aerogels (as drug carriers) and evaluation of mathematicalkinetic release models, Iran J. Pharm. Res. 17(2018) 818-829.

[51]   M. Kéri, A. Forgács, V. Papp, I. Bányai, P. Veres, A. Len, Z. Dudás, I. Fábián, J. Kalmár, Gelatin content governs hydration induced structural changes in silica-gelatin hybrid aerogels - Implications in drug delivery, Acta Biomater. 105(2020) 131-145. https://doi.org/10.1016/j.actbio.2020.01.016

[52]   G. Nagy, G. Király, P. Veres, I. Lázár, I. Fábián, G. Bánfalvi, I. Juhász, J. Kalmár, Controlled release of methotrexate from functionalized silica-gelatin aerogel microparticles applied against tumor cell growth, Int. J. Pharm. 558(2019) 396-403. https://doi.org/10.1016/j.ijpharm.2019.01.024

[53]   S. Ye, S. He, C. Su, L. Jiang, Y. Wen, Z. Zhu, W. Shao, Morphological, release and antibacterial performances of amoxicillin-loaded cellulose aerogels, Molecules. 23(2018) pii: E2082. https://doi.org/10.3390/molecules23082082

[54]   C. López-Iglesias, J. Barros, I. Ardao, P. Gurikov, F.J. Monteiro, I. Smirnova,C.Alvarez-Lorenzo, C.A. García-González, Jet cutting technique for the production of chitosan aerogel microparticles loaded with vancomycin, Polymers 12(2020) 273. https://doi.org/10.3390/polym12020273

[55]   F. Sabri, J.A. Cole, M.C. Scarbrough, N. Leventis, Investigation of polyurea-crosslinked silica aerogels as a neuronal scaffold: a pilot study, PLoS One. 7(2012) e33242. https://doi.org/10.1371/journal.pone.0033242

[56]   F. Sabri, J.D. Boughter Jr, D. Gerth, O. Skalli, T-C.N. Phung, G-R.M. Tamula, N. Leventis, Histological evaluation of the biocompatibility of polyurea crosslinked silica aerogel implants in a rat model: a pilot study. PLoS One. 7(2012) e50686. https://doi.org/10.1371/journal.pone.0050686

[57]   F. Sabri, D. Gerth, G.R.Tamula, T.C. Phung, K.J. Lynch, J.D. BoughterJr, Noveltechnique for repair of severed peripheral nerves in rats using polyurea crosslinked silicaaerogel scaffold, J. Invest. Surg. 27(2014) 294-303. https://doi.org/10.3109/08941939.2014.906688

[58]   K.J. Lynch, O. Skalli, F. Sabri, Investigation of surface topography and stiffness on adhesion and neurites extension of PC12 cells on crosslinked silica aerogel substrates, PLoS One. 12(2017) e0185978. https://doi.org/10.1371/journal.pone.0185978

[59]   W. Yin,S.M. Venkitachalam, E. Jarrett, S. Staggs, N. Leventis, H. Lu, D.A. Rubenstein, Biocompatibility of surfactant-templated polyurea-nanoencapsulated macroporous silica aerogels with plasma platelets and endothelial cells, J. Biomed. Mater. Res. A. 92 (2010) 1431-1439. https://doi.org/10.1002/jbm.a.32476

[60]   W. Yin, H. Lu, N. Leventis, D.A. Rubenstein, Characterization of the biocompatibility and mechanical properties of polyurea organic aerogels with the vascular system: potential as a blood implantable material, Int. J. Polym. Mater. 62 (2013) 109-118. https://doi.org/10.1080/00914037.2012.698339

[61]   R. Ghafari, M. Jonoobi, L.M. Amirabad, K. Oksman, A.R. Taheri, Fabrication and characterization of novel bilayer scaffold from nanocellulose based aerogel for skin tissueengineering applications. Int. J. Biol. Macromol. 136 (2019) 796-803. https://doi.org/10.1016/j.ijbiomac.2019.06.104

[62]   R.R. Mallepally, M.A. Marin, V. Surampudi, B. Subia, R.R. Rao, S.C. Kundu, M.A. McHugh, Silk fibroin aerogels: potential scaffolds for tissue engineering applications. Biomed Mater. 10 (2015) 035002. https://doi.org/10.1088/1748-6041/10/3/035002

[63]   H. Maleki, M.A. Shahbazi, S. Montes, S.H. Hosseini, M.R. Eskandari, S. Zaunschirm,T. Verwanger, S. Mathur, B. Milow, B. Krammer, N. Hüsing, Mechanically strong silica-silk fibroin bioaerogel: a hybrid scaffold with ordered honeycomb micromorphology and multiscale porosity for bone regeneration,ACS Appl. Mater. Interfaces. 11(2019) 17256-17269. https://doi.org/10.1021/acsami.9b04283

[64]   D. Sun, W. Liu, A. Tang, F. Guo, W. Xie, A new PEGDA/CNF aerogel-wet hydrogel scaffold fabricated by a two-step method, Soft Matter. 15(2019) 8092-8101. https://doi.org/10.1039/C9SM00899C

[65]   A. Tang, J. Li,J. Li, S. Zhao, W. Liu, T. Liu, J. Wang, Y. Liu, Nanocellulose/PEGDA aerogel scaffolds with tunable modulus prepared by stereolithography for three-dimensional cell culture, J. Biomater. Sci. Polym. Ed. 30(2019) 797-814. https://doi.org/10.1080/09205063.2019.1602904

[66]   J. Liu, F. Cheng, H. Grénman, S. Spoljaric, J. Seppälä, J.E. Eriksson, S. Willför, C. Xu, Development of nanocellulose scaffolds with tunable structures to support 3D cell culture, CarbohydrPolym. 148(2016) 259-271. https://doi.org/10.1016/j.carbpol.2016.04.064

[67]   D.A. Rubenstein, H. Lu, S.S. Mahadik, N. Leventis, W. Yin, Characterization of the physical properties and biocompatibility of polybenzoxazine-based aerogels for use as anovel hard-tissue scaffold, J. Biomater. Sci. Polym. Ed. 23(2012) 1171-1184. https://doi.org/10.1163/092050611X576954

[68]   J. Liao, Y. Qu, B. Chu, X. Zhang, Z. Qian, Biodegradable CSMA/PECA/Graphene porous hybrid scaffold for cartilage tissue engineering. Sci. Rep. 5(2015) 9879. https://doi.org/10.1038/srep09879

[69]   T. Hu, J. Xu, Y. Ye, Y. Han, X. Li, Z. Wang, D. Sun, Y. Zhou, Z. Ni, Visual detection of mixed organophosphorous pesticide using QD-AChE aerogel based microfluidic arrays sensor, Biosens. Bioelectron. 136 (2019) 112-117. https://doi.org/10.1016/j.bios.2019.04.036

[70]   J.M. Jeong, M. Yang, D.S. Kim, T.J. Lee, B.G. Choi, D.H. Kim, High performance electrochemical glucose sensor based on three-dimensional MoS_2/graphene aerogel, J. Colloid Interface Sci. 506 (2017) 379-385. https://doi.org/10.1016/j.jcis.2017.07.061

[71]   L. Ruiyi, C. Fangchao, Z. Haiyan, S. Xiulan, L. Zaijun, Electrochemical sensor for detection of cancer cell based on folic acid and octadecylamine-functionalized graphene aerogel microspheres, Biosens. Bioelectron. 119(2018) 156-162. https://doi.org/10.1016/j.bios.2018.07.060

[72]   R. Li, T. Yang, Z. Li, Z. Gu, G. Wang, J. Liu, Synthesis of palladium@gold nanoalloys/nitrogen and sulphur-functionalized multiple graphene aerogel for electrochemical detection of dopamine, Anal. Chim. Acta. 954(2017) 43-51. https://doi.org/10.1016/j.aca.2016.12.031

[73]   S. Dong, N. Li, G. Suo, T. Huang, Inorganic/organic doped carbon aerogels as biosensing materials for the detection of hydrogen peroxide, Anal Chem. 85 (2013) 11739-11746. https://doi.org/10.1021/ac4015098

[74]   B. Wang, S.Yan, Direct electrochemical analysis of glucose oxidase on a graphene aerogel/gold nanoparticle hybrid for glucose biosensing, J. Solid State Electrochem. 19(2015) 307-314. https://doi.org/10.1007/s10008-014-2608-7

[75]   Y.Gao, F.Yang, Q. Yu, R. Fan, M. Yang, S. Rao, Q. Lan, Z. Yang, Z. Yang, Three-dimensional porous Cu@Cu$_2$O aerogels for direct voltammetric sensing of glucose, Mikrochim Acta. 186(2019) 192. https://doi.org/10.1007/s00604-019-3263-6

[76]   Y. Wu, L. Jiao, W. Xu, W. Gu, C. Zhu, D. Du, Y. Lin, Polydopamine-capped bimetallic AuPt hydrogels enable robust biosensor for organophosphorus pesticide detection, Small. 15(2019) e1900632. https://doi.org/10.1002/smll.201900632

Materials Research Forum LLC
https://doi.org/10.21741/9781644901298-3

Chapter 3

Bioaerogels: Synthesis Approaches, Biomedical Applications and Cell Uptake

Jhonatas Rodrigues Barbosa[1*], Rafael Henrique Holanda Pinto[1], Luiza Helena da Silva Martins[2], Raul Nunes de Carvalho Junior[1,3]

[1]LABEX/FEA (Faculty of Food Engineering), Program of Post-Graduation in Food Science and Technology, Federal University of Pará, Rua Augusto Corrêa S/N, Guamá, 66075-900 Belém, Pará, Brazil

[2]Institute of Animal Health and Production - Federal Rural University of Amazonia - Avenida Presidente Tancredo NevesNº 2501; Terra Firme; Cep: 66.077-830Belém, Pará, Brazil

[3]Program of Post-Graduation in Natural Resources Engineering, Federal University of Pará, Rua Augusto Corrêa S/N, Guamá, 66075-900 Belém, Pará, Brazil

jhonquimbarbosa@gmail.com; Orcid iD: https://orcid.org/0000-0002-6394-299X

Abstract

Bioaerogels are a special class of aerogels, produced from natural polymers, are porous structures with promising physicochemical properties for various applications. This chapter focus on the latest bioaerogel findings, addressing the synthesis, impregnation of bioactive compounds, pharmacological applications and aspects of cell uptake, biodegradability and toxicity. Bioaerogels are biomaterials with interesting properties such as high surface area, high thermal and mechanical resistance, low density and dielectric constant. It has been reported that these biomaterials can be used for drug delivery and molecular scaffolding production. Furthermore, it has been shown that bioaerogels are biocompatible, biodegradable, non-toxic, and can be absorbed and degraded by the cellular environment. Finally, bioaerogels are promising, inexpensive, environmentally friendly and versatile materials and can be the basis for the manufacture of new technologies and biomaterials.

Keywords

Biopolymer, Bioaerogel, Biomaterial, Biodegradability, Cellular Uptake

Contents

1.   Introduction

Bioaerogels are organic biomaterials, polymeric, produced from natural sources, porous, with high surface area, high thermal and mechanical resistance, low density, and dielectric constant. These properties make interesting biomaterial bioaerogels to be used as carriers for the delivery of bioactive drugs and compounds, production of tissue and bone engineering mulches, biosensors among other applications [1,2].

Traditional aerogels made from inorganic materials and synthetic polymers go through the following steps: gel synthesis, gel alcohol formation and finally, by drying techniques. Bioaerogels are produced by the same techniques, however, the use of natural polymers such as polysaccharides and application of bioactive compounds makes these biomaterials known as bioaerogels [3]. Bioaerogels are mesoporous, forming compact structures of polymers linked by strong molecular entanglements. The structures of these biopolymers are resistant to 90% of voids and can be used for various applications [4].

Although some materials are used for the production of aerogels such as silica, metal oxides, carbon allotropes, and transition metals, natural polymers such as pectin, chitosan, cellulose, studied form a new class of biomaterial, recognized as bioaerogel. Bioaerogels are more interesting when compared to traditional aerogels for several reasons such as natural polymers are abundant and cheap, bioaerogels production is environmentally friendly, disposal of these biomaterials is not harmful to the environment and ultimately bioaerogels are biocompatible, biodegradable and non-toxic to the cellular environment [5,6].

Bioaerogels have been applied in several areas, especially in the pharmaceutical industry [7]. The use of bioaerogels for drug delivery and bioactive compounds has been reported, as Mustapa et al. [8], alginate and silica bioaerogel were used to impregnate herbal extracts rich in phytol. The alginate bioaerogel presented the largest load impregnated with phyto land the best release kinetics when compared to silica air gel. Other studies show that bioaerogels have good applicability as insulators, catalysts or catalyst support [9].

This chapter presents the main research with emphasis on bioaerogel structures, synthesis and processing methods, biomedical applications and properties. Biomedical applications addressed the use of bioaerogels in drug delivery and tissue engineering scaffolding. Finally, aspects related to biocompatibility, biodegradability, toxicity, absorption and cellular degradation would be clarified.

2. Bioaerogels synthesis

2.1 Chitin bioaerogels

The synthesis of bioaerogels comprises a set of basic techniques such as gelatinization, alcohol formation and finally drying, which may be by lyophilization and supercritical drying (Fig. 1). Tsioptias et al. [10] synthesized chitin bioaerogel using supercritical fluid technology. The sample was solubilized in a solution of dimethylacetamide and LiCl. The biomaterial formed was molded and washed with water at room temperature to form the hydrogel. The conversion of hydrogel to alcogel occurred using methanol and propanol. The drying of the alcogel was carried out using supercritical CO_2 at temperature and pressure levels that varied from 40-80°C and 80-300 bar, respectively. The mass flow rate of CO_2 was 0.42 g.min^{-1} over 2 hours, forming chitin bioaerogel.

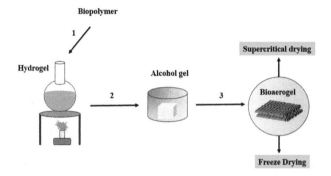

Fig. 1 *Graphic summary of bioaerogel synthesis. Step (1), gelatinization of the polymers. Step (2), alcohol formation. Step (3), lyophilization and supercritical drying.*

Silva et al. [11] synthesized bioaerogels of ionic solubilized chitin (1-butyl-3-imidazolium acetate) and drying was performed by supercritical CO_2. The study showed that the synthesized bioaerogel presents low toxicity levels, indicating that they are materials with potential biomedical applications as tissue repair. In a similar study Dassanayake and co-workers [12] synthesized chitin aerogel from shrimp shells solubilized in sodium hydroxide and urea-water solution. The developed activated carbon was used as an adsorbent and was applied to the chitin bioaerogels for CO_2 recovery.

2.2 Chitosan bioaerogels

Robtizer and co-workers [13] synthesized chitosan bioaerogels with liquid CO_2 and supercritical CO_2. The hydrogel was formed after dissolution of chitosan in aqueous acetic acid solution. After the addition of drops of the polymer in sodium hydroxide solution, pH inversion was observed. Hydrogel dehydration occurred by immersion in ethanol-water solutions with alcohol concentrations in (10, 30, 50, 70, 90 and 100%) for 15 minutes. Ethanol was replaced by super critical CO_2 and bioaerogels were obtained by drying with supercritical CO_2 forming microspheres. The synthesis method allowed the maintenance of the chitosan structure, as well as a satisfactory shrinkage.

Singh et al. [14] developed chitosan bioaerogels with supercritical fluid from chitosan-L-glutamic acid. Hydrogel derivatives were prepared from chitosan powder, dissolved in aqueous acetic acid solution. The conversion of hydrogel to alcogel was performed with acetone and ethanol. The liquid organic solvents were removed with supercritical CO_2 under the following conditions: temperature of 40 ° C and pressure of 100-200 bar for 90

Aerogels II: Preparation, Properties and Applications   Materials Research Forum LLC
Materials Research Foundations **97** (2021) 43-56   https://doi.org/10.21741/9781644901298-3

minutes. The release of CO_2 at the end of the process allowed the formation of bluish-colored pulverized material.

Investigations by Radwan-Praglowska et al. [15] showed the synthesis of aerogel from the dissolution of chitosan in a mixture containing extracts of Tiliaplatyphyllos in acetic acid solution, addition of aspartic, glutamic acid and propylene glycol. The mixture was placed in a microwave and further washed, followed by freeze-drying. This mechanism contributed to the production of bioaerogels impregnated with bioactive compounds and swelling properties.

2.3   Alginate and agar bioaerogels

Wang et al. [16] synthesized calcium alginate bioaerogels from sodium alginate and $CaCl_2$ solution. The initially formed hydrogel was frozen and lyophilized for 12 hours. The synthesized product was applied as adsorbent to remove Pb from wastewater. This type of aerogel is environmentally friendly and cost effective. The study results showed that this type of material allows the adsorption of up to 96% Pb in aqueous solution. The regeneration of this type of aerogel can occur through simple mechanisms, with acid use, without loss of performance.

Pantic co-workers [17] developed supercritical CO_2 alginate aerogel. Sodium salt and alginic acid were solubilized in water and $CaCl_2$ solution added dropwise. The hydrogel formed was converted to alcohol-gel by immersion in ethanol: The drying of the alcohol balls were performed under the following supercritical conditions: 40 °C, 100-150 bar, with total ethanol removal time of 6 hours.

The research by Paraskevopoulou et al. [18] developed tungsten alginate bioaerogel through a mixture of sodium alginate and $CaCO_3$ solution in water, with the addition of δ-gluconolactone. The mixture was kept refrigerated to complete the gelling process. The gels were treated with ethanol-water, pure ethanol, and dried with supercritical CO_2. This type of aerogel can be applied in catalysis processes. Li et al. [19] synthesized alginate bioaerogel using clay, aqueous sodium salt solution and drops of p-toluenesulfonic acid monohydrate (pH 6-8). The mixture obtained was lyophilized to obtain alginate aerogel. Tan et al. [20] produced graphite-agar carbon nitride aerogel. The resulting material presented satisfactory 3D structure. The temperature level applied for agar gel formation was 85°C. The gel produced was dried in supercritical fluid.

2.4   Cellulose bioaerogels

Zeng and co-workers [21] obtained cellulose aerogel from waste denim, cotton, and microcrystalline cotton, using ionic solvents through dissolution, regeneration, and

Aerogels II: Preparation, Properties and Applications   Materials Research Forum LLC
Materials Research Foundations **97** (2021) 43-56   https://doi.org/10.21741/9781644901298-3

drying. Hydrogel formation occurred through the process of dissolving in ionic liquids at 100°C. Dehydration of the hydrogel occurred through lyophilization and supercritical fluid methods. Sehaqui and co-workers [22] synthesized low-density bioaerogels from aqueous nanofibrilated cellulose dispersions using different solvents. In this study, nanofiber dispersion was preserved in aerogel.

3.   Biomedical applications of bioaerogels

3.1   Bioaerogels can be used to support drug administration

Bioaerogels have physicochemical and biological properties of interest such as satisfactory biocompatibility, high internal surface areas, and surface/volume ratios, favoring drug loading. For this reason, they are characterized as potential biomaterials for drug transport. Several methods have been developed to transport bioactive compounds in bioaerogels, such as adding the drugs to the mixture before the gelation step (hydrogel formation), addition of the drug during the aging step and adding drugs during the adsorption / sedimentation steps in the produced aerogel [23]. The graphical summary of the application of bioaerogels as platforms for controlled drug release is shown in (Fig. 2).

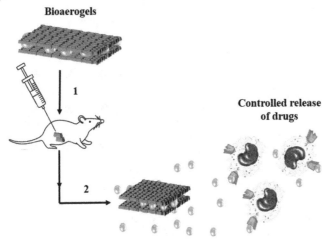

Fig. 2 *Graphic summary of the application of bioaerogels as platforms for controlled drug release. Step (1), Application of drug-loaded bioaerogels in rats. Step (2), controlled release of the drug into the cellular environment.*

Polysaccharide bioaerogel particles are potential drug carriers for the pulmonary system contributing to better airflow. Aerogel surface area and volume are parameters that determine drug control, and release. Drug loading can be increased by increasing the porosity of the encapsulating material. Larger surface carriers allow increased diffusion and solubility of drugs. However, it is important to investigate the relationship between the release profile of bioactive compounds and the structure of bioaerogels [23].

Examples of bioaerogel applications in drug delivery include alginate aerogel in the transport of anti-inflammatory, and non-steroids [24]; chitosan aerogels for *in vitro* administration of ketoprofen and benzoic acid [25] and cellulose nanofiber aerogels for fluorouracil administration [26].

3.2   Bioaerogels can be used as tissue engineering scaffolding

Current scientific research in tissue engineering was developed with the aim of synthesizing 3D scaffolding with cell-forming properties capable of restructuring tissues and organs of the human body. The properties of bioaerogel structures are important when applied to tissue regeneration technology because through the cellular connection in pores, they provide the necessary energy supply for the elimination of metabolic byproducts [27]. Graphical summary of the application of bioaerogels used as tissue engineering scaffolds can be observed (Fig.3).

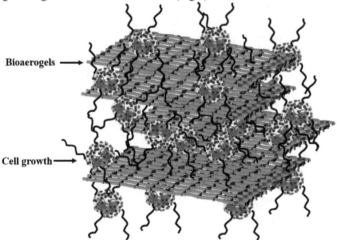

Fig. 3 Graphic summary of the application of bioaerogels used as tissue engineering structures. Bioaerogels serve as scaffolding for cells to grow and develop.

Aerogels II: Preparation, Properties and Applications   Materials Research Forum LLC
Materials Research Foundations 97 (2021) 43-56   https://doi.org/10.21741/9781644901298-3

Alginate and nanocellulose bioaerogels are commonly applied in tissue engineering for presenting better cost benefit, low level of toxicity and easy gelation process (hydrogel formation).However, adsorption on protein alginate surfaces is limited due to hydrophilic properties that prevent cell adhesion. Thus, its applications are restricted in tissue engineering. Applications of tissue engineering bioaerogels include scaffolding to support the growth and formation of tissue cells (*in vitro*) from nanocellulose bioaerogels and biological dressings for cell proliferation [28].

4.   Biocompatibility, toxicity, biodegradability and intracellular absorption

4.1   Bioaerogels have biocompatibility and low toxicity

Biocompatibility is a pharmacological term used to determine whether a substance is biocompatible with the cellular environment and whether cellular receptors recognize a substance as pathogenic or not. Biocompatibility is an important pharmacological characterization text, a prerequisite for the production of any new drugs [29]. Several *in vitro* and *in vivo* texts have already been applied to evaluate the biocompatibility of various materials, chemical substances, synthetic polymers, and biopolymers. Depending on the type of biopolymers, some techniques of biocompatibility analysis, biodegradation, and metabolic fate are explored. For example, chitosan is one of the most commonly used natural biopolymers as a polymeric matrix for delivery, drug loading, and bioactive compounds. In this particular case, the main technique used is *in vitro* text, where precultured cells in suitable media are exposed to the substances of interest for periods of up to 96 h [29].

In the work described by Onishi and Machida [30], mice were used to evaluate the effects of 50% deacetylated chitosan biodegradability. In *in vivo*, a solution containing deacetylated chitosan was applied via intraperitoneal and the chitosan corporation distribution was evaluated. The results showed that chitosan did not accumulate between the organs and was excreted by the urinary system, being therefore, biocompatible with the cellular system. In the *in vitro* study, incubation models with lysozyme enzyme were explored, it was found that the rate of biopolymer degradation was accelerated, indicating that chitosan can be applied in food formulations and used for delivery of drugs and bioactive compounds. Rao and Sharma [31] studied Chitosan toxicity in *in vitro* and *in vivo* models with mouse models and no significant toxicity was noted.

Fernández-Cossío and co-workers [32], *in vivo* studies, evaluated the bioactivity and biocompatibility of the toxicity of agarose gel as a tissue filler in hepidermal implants. The results showed that there were no side effects, region or even infections. Furthermore, it was observed that macrophages were responsible for implant degradation, and based on histopalogical results no necrosis, calcification, tumor formation or even gel migration were observed for an 8-month evaluation period.

The biocompatibility of alginate microcapsules was evaluated by several in vitro textures Orive et al. [33]. Degree of purity, viscosity, and content of polyphenols and proteins were evaluated using sensitive techniques, and their influence was addressed. Based on the results, the authors conclude that alginates can be used as a biocompatible polymeric matrix for the transport and implantation of drugs and bioactive compounds, as well as for cell immobilization. Tam et al. [34] found that molecular details of alginates and their gels may influence biocompatibility and toxicity. The contents of manuronate (M) and guluronate (G) were found to affect the solubility, viscosity and therefore, biocompatibility of this polymer in the cellular environment. The authors concluded that alginate with higher viscosity, therefore, intermediate content of guluronate (G) are more biocompatible.

4.2 Bioaerogels are biodegradable and are absorbed via intracellular

Bioaerogels can be absorbed via the intracellular route, where they are recognized by several cell receptors. Cellular receptors are distributed throughout the cell membrane and are available to be activated. Among these receptors, some are quite important, such as Toll-like receptors (TLRS) and mannose receptors [35]. According, Rodríguez-jorge et al. [36], other receptors may be involved in the absorption of substances in the cellular environment, such as receptors present in the bloodstream, such as serum proteins and phytolites. In addition, other binding molecules associated with lipopolysaccharides (LBP), and T cell antigen receptors can be activated [37].

Miyamoto et al. [38], applying *in vivo* methods, evaluated the absorption and reaction of living tissue from polymeric matrices of cellulose and derivatives such as ethyl cellulose, hydroxyethyl cellulose, aminoethyl cellulose and cellulosic polyonic complexes. Absorption through cell tissue was controlled by the basic macrophage mechanisms. However, it depended on certain intrinsic characteristics of the biopolymer such as crystallinity, structure, and molecular conformation, while the reaction to foreign body text was considered moderate, the same for all textures. The authors consider that cellulose and its derivatives are biocompatible biomaterials that can be absorbed, metabolized and biodegraded by the cellular environment.

Studies on chitosan uptake and cell distribution have not yet been reported, however, chitosan complexes with other polymers have been reported. Chitosan has been reported to penetrate the cell membrane through electron disturbances caused by cationic charge [39]. It is known that temperature may influence intracellular absorption of chitosan complexes, temperatures at 37 °C were 3 times more efficient than at 4 °C, probably some non-Adenosine triphosphate (ATP) pathway endocytic mechanism may be involved [39]. Naeiniet al [40], demonstrated that the chitosan and deoxyribonucleic acid (DNA) complex particles are observed by the cells and part of this complex remains in the cell cytosol, probably some unidentified endocytic mechanism may be responsible for the absorption process.

Conclusions

The present chapter reveals that bioaerogels have interesting properties;also, the use of natural polymers has attracted the attention of investors, due to the qualities of these biomaterials, such as biodegradability, biocompatibility and abundance of these polymers in nature. The bioaerogel synthesis process has been shown to be similar to the manufacture of silica aerogels and other inorganic materials. It has been reported that the stages of the formation of sol-gel, alcohol gel, and finally aerogels are similar. However, the use of natural polymers and the impregnation of bioactive compounds leads to the differentiation of so-called bioaerogels. Finally, the drying of the bioaerogel, by lyophilization, supercritical or even by evaporation, influences the properties of bioaerogels. Several biomedical applications of bioaerogels have been reported; showing that bioarogels are safe, biodegradable, biocompatible, and non-toxic to the cellular environment. It has been reported that aerogels can be absorbed and metabolized by the intracellular environment. In short, bioaerogels are promising biomaterials for the production of new, inexpensive, efficient, and differentiated technologies. Additional studies are still needed, however, based on current results, expectations for future applications of these biomaterials suggest that they may be the basis for new biomaterials.

Acknowledgment

The author's acknowledgment CNPq (National Council for Scientific and Technological Development), process number 169983/2018-8, scholarship grant and UFPA (Federal University of Pará), Brazil, for the space of development and scientific research.

References

[1] H. Maleki, L. Durães, C.A. García-González, P. del Gaudio, A. Portugal, M. Mahmoudi, Synthesis and biomedical applications of aerogels: possibilities and challenges, Adv. Colloid Interface Sci. 236 (2016) 1-27. https://doi.org/10.1016/j.cis.2016.05.011

[2] J. Alemán, A.V. Chadwick, J. He, M. Hess, K. Horie, R.G. Jones, P. Kratochvíl, I. Meisel, I. Mita, G. Moad, Definitions of terms relating to the structure and processing of sols, gels, networks, and inorganic-organic hybrid materials (IUPAC Recommendations 2007), Pure Appl. Chem. 79 (2007) 1801-1829. https://doi.org/10.1351/pac200779101801

[3] H. Maleki, L. Durães, A. Portugal, An overview on silica aerogels synthesis and different mechanical reinforcing strategies, J. Non-Cryst. Solids. 385 (2014) 55-74. https://doi.org/ 10.1016/j.jnoncrysol.2013.10.017

[4] I.M. El-Nahhal, N.M. El-Ashgar, A review on polysiloxane-immobilized ligand systems: synthesis, characterization and applications, J. Organomet. Chem. 692 (2007) 2861-2886. https://doi.org/10.1016/j.jorganchem.2007.03.009

[5] A.C. Pierre, G.M. Pajonk, Chemistry of aerogels and their applications, Chem. Rev. 102 (2002) 4243-4266. https://doi.org/10.1021/cr0101306

[6] S.W. Ruban, Biobased packaging-application in meat industry, Vet. World. 2 (2009) 79-82. https://doi.org/10.1016/j.procbio.2015.02.009

[7] H. Derakhshankhah, M.J. Hajipour, E. Barzegari, A. Lotfabadi, M. Ferdousi, A.A. Saboury, E.P. Ng, M. Raoufi, H. Awala, S. Mintova, Zeolite nanoparticles inhibit Aβ– fibrinogen interaction and formation of a consequent abnormal structural clot, ACS Appl. Mater. Interfaces. 8 (2016) 30768-30779. https://doi.org/ 10.1021/acsami.6b10941

[8] A. Mustapa, A. Martin, L. Sanz-Moral, M. Rueda, M. Cocero, Impregnation of medicinal plant phytochemical compounds into silica and alginate aerogels, J. Supercrit. Fluids. 116 (2016) 251-263. https://doi.org/10.1016/j.supflu.2016.06.002

[9] B. Ding, J. Cai, J. Huang, L. Zhang, Y. Chen, X. Shi, Y. Du, S. Kuga, Facile preparation of robust and biocompatible chitin aerogels, J. Mater. Chem. 22 (2012) 5801-5809. https://doi.org/10.1039/c2jm16032c

[10] C. Tsioptsias, C. Michailof, G. Stauropoulos, C. Panayiotou, Chitin and carbon aerogels from chitin alcogels, Carbohydr. Polym. 76 (2009) 535–540. https://doi.org/10.1016/j.carbpol.2008.11.018

[11] S.S. Silva, A.R.C. Duarte, A.P. Carvalho, J.F. Mano, R.L. Reis, Green processing of porous chitin structures for biomedical applications combining ionic liquids and supercritical fluid technology, Acta Biomater. 7 (2011) 1166–1172. https://doi.org/10.1016/j.actbio.2010.09.041

[12] R.S. Dassanayake, C. Gunathilake, N. Abidi, M. Jaroniec, Activated carbon derived from chitin aerogels: preparation and CO_2 adsorption, Cellulose 25 (2018) 1911–1920. https://doi.org/10.1007/s10570-018-1660-3

[13] M. Robitzer, F. D. Renzo, F. Quignard, Natural materials with high surface area. Physisorption methods for the characterization of the texture and surface of polysaccharide aerogels, Microporous Mesoporous Mater. 140 (2011) 9–16. https://doi.org/10.1016/j.micromeso.2010.10.006

[14] J. Singh, P. Dutta, J. Dutta, A. Hunt, D. Macquarrie, J. Clark, Preparation and properties of highly soluble chitosan–l-glutamic acid aerogel derivative, Carbohydr. Polym. 76 (2009) 188–195. https://doi.org/10.1016/j.carbpol.2008.10.011

[15] J. Radwan-Pragłowska, M. Piątkowski, Ł. Janus, D. Bogdał, D. Matysek, V. Cablik, Microwave-assisted synthesis and characterization of antioxidant chitosan-based aerogels for biomedical applications, Int. J. Polym. Anal. Charact. 1 (2018) 1–9. https://doi.org/10.1080/1023666X.2018.1504471

[16] Z. Wang, Y. Huang, M. Wang, G. Wu, T. Geng, Y. Zhao, A. Wu, Macroporous calcium alginate aerogel as sorbent for Pb^{2+} removal from water media, J. Environ. Chem. Eng. 4 (2016) 3185–3192. https://doi.org/10.1016/j.jece.2016.06.032

[17] M. Pantić, Ž. Knez, Z. Novak, Supercritical impregnation as a feasible technique for entrapment of fat-soluble vitamins into alginate aerogels, J. Non-Cryst. Solids. 432 (2016) 519–526. https://doi.org/10.1016/j.jnoncrysol.2015.11.011

[18] P. Paraskevopoulou, P. Gurikov, G. Raptopoulos, D. Chriti, M. Papastergiou, Z. Kypritidou, V. Skounakis, A. Argyraki, Strategies toward catalytic biopolymers: incorporation of tungsten in alginate aerogels, Polyhedron. 154 (2018) 209–216. https://doi.org/10.1016/j.poly.2018.07.051.

[19] X.L. Li, M.J. Chen, H.B. Chen, Facile fabrication of mechanically-strong and flame retardant alginate/clay aerogels, Compos. Part B Eng. 164 (2019) 18–25. https://doi.org/10.1016/j.compositesb.2018.11.055.

[20] L. Tan, C. Yu, M. Wang, S. Zhang, J. Sun, S. Dong, J. Sun, Synergistic effect of adsorption and photocatalysis of 3D g-C3N4-agar hybrid aerogels, Appl. Surf. Sci. 467 (2019) 286–292. https://doi.org/10.1016/j.apsusc.2018.10.067

[21] B. Zeng, X. Wang, N. Byrne, Development of cellulose based aerogel utilizing waste denima morphology study, Carbohydr. Polym. 205 (2019) 1–7. https://doi.org/10.1016/j.carbpol.2018.09.070

Materials Research Forum LLC
https://doi.org/10.21741/9781644901298-3

[22] H. Sehaqui, Q. Zhou, L.A. Berglund, High-porosity aerogels of high specific surface area prepared from nanofibrillated cellulose (NFC), Compos. Sci. Technol. 71 (2011) 1593–1599. https://doi.org/10.1016/j.compscitech.2011.07.003

[23] F. P. Soorbaghi, M. Isanejad, S. Salatin, M. Ghorbani, S. Jafari, H. Derakhshankhah, Bioaerogels: Synthesis approaches, cellular uptake, and the biomedical applications, Biomed. Pharmacother. 111 (2019) 964–975.
https://doi.org/10.1016/j.biopha.2019.01.014

[24] R.H. Schmedlen, K.S. Masters, J.L. West, Photocrosslinkable polyvinyl alcohol hydrogels that can be modified with cell adhesion peptides for use in tissue engineering, Biomaterials. 23 (2002) 4325–4332. https://doi.org/10.1016/S0142-9612(02)00177-1

[25] D. Nguyen, D.A. Hägg, A. Forsman, J. Ekholm, P. Nimkingratana, C. Brantsing, T. Kalogeropoulos, S. Zaunz, S. Concaro, M. Brittberg, Cartilage tissue engineering by the 3D bioprinting of iPS cells in a nanocellulose/alginate bioink, Sci. Rep. 7 (2017) 658. https://doi.org/10.1007/s12325-016-0470-y

[26] K.B. Fonseca, S.J. Bidarra, M.J. Oliveira, P.L. Granja, C.C. Barrias, Molecularly designed alginate hydrogels susceptible to local proteolysis as three-dimensional cellular microenvironments, Acta Biomater. 7 (2011) 1674–1682.
https://doi.org/10.1016/j.actbio.2010.12.029

[27] Y.Y. Jo, S.G. Kim, K.J. Kwon, H. Kweon, W.S. Chae, W.G. Yang, E.Y. Lee, H. Seok, Silk fibroin-alginate-hydroxyapatite composite particles in bone tissue engineering applications *in vivo*, Int. J. Mol. Sci. 18 (2017) 858.
https://doi.org/10.3390/ijms18040858.

[28] S. Frindy, A. Primo, H. Ennajih, A. el Kacem Qaiss, R. Bouhfid, M. Lahcini, E.M. Essassi, H. Garcia, A. El Kadib, Chitosan–graphene oxide films and CO_2-dried porous aerogel microspheres: interfacial interplay and stability, Carbohydr. Polym. 167 (2017) 297–305. https://doi.org/10.1016/j.carbpol.2017.03.034

[29] T. Kean, M. Thanou, Biodegradation, biodistribution and toxicity of chitosan, Adv. Drug Deliv. Rev. 62 (2010) 3-11. https://doi.org/10.1016/j.addr.2009.09.004

[30] H. Onishi, Y. Machida, Biodegradation and distribution of water-soluble chitosan in mice, Biomaterials. 20 (1999) 175-182. https://doi.org/10.1016/S0142-9612(98)00159-8

[31] S.B. Rao, C.P. Sharma, Use of chitosan as a biomaterial: studies on its safety and hemostatic potential, J. Biomed. Mater. Res. 34 (1997) 21-28. https://doi.org/10.1002/(SICI)1097-4636(199701)34:1<21::AID-JBM4>3.0.CO;2-P.

[32]S. Fernández-Cossío, A. León-Mateos, F.G. Sampedro, M.T.C. Oreja, Biocompatibility of agarose gel as a dermal filler: histologic evaluation of subcutaneous implants, Plast. Reconstr. Surg. 120 (2007) 1161-1169. https://doi.org/10.1097/01.prs.0000279475.99934.71

[33] G. Orive, A. Carcaboso, R. Hernandez, A. Gascon, J. Pedraz, Biocompatibility evaluation of different alginates and alginate-based microcapsules, Biomacromolecules. 6 (2005) 927-931. https://doi.org/10.1021/bm049380x

[34] S.K. Tam, J. Dusseault, S. Bilodeau, G. Langlois, J.P. Hallé, L.H. Yahia, Factors influencing alginate gel biocompatibility, J. Biomed. Mater. Res. Part A. 98 (2011) 40-52. https://doi.org/10.1002/jbm.a.33047

[35] J.R. Barbosa, M.M.S. Freitas, L.H.S. Martins, R.N.C. Junior, Polysaccharides of mushroom *Pleurotus spp*: New extraction techniques, biological activities and development of new technologies, Carbohydr. Polym. 229 (2019)115550. https://doi.org/10.1016/j.carbpol.2019.115550

[36] O. Rodríguez-jorge, L.A. Kempis-calanis, W. Abou-jaoudé, D.Y. Gutiérrez-reyna, C. Hernandez, O. Ramirez-pliego, Cooperation between T cell receptor and Toll-like receptor 5 signaling for CD4$^+$ T cell activation, Sci. Signal. 88 (2019) 1–11. https://doi.org/10.1126/scisignal.aar3641

[37] F.A.M. Saner, A. Herschtal, B.H. Nelson, E.L. Goode, S.J. Ramus, A. Pandey, D.D.L. Bowtell, in patients with cancer. Nat. Rev. Cancer. 19 (2019) 339–348. https://doi.org/10.1038/s41568-019-0145-5

[38] T. Miyamoto, Si. Takahashi, H. Ito, H. Inagaki, Y. Noishiki, Tissue biocompatibility of cellulose and its derivatives, J. Biomed. Mater. Res. 23 (1989) 125-133. https://doi.org/10.1002/jbm.820230110

[39] P.L. Ma, M. Lavertu, F.M. Winnik, M.D. Buschmann, Stability and binding affinity of DNA/chitosan complexes by polyanion competition, Carbohydr. Polym. 176 (2017) 167-176. https://doi.org/10.1016/j.carbpol.2017.08.002

[40] A.T. Naeini, O.Y. Soliman, M.G. Alameh, M. Lavertu, M.D. Buschmann, Automated in-line mixing system for large scale production of chitosan-based polyplexes, J. Colloid Interface Sci. 500 (2017) 253-263. https://doi.org/10.1016/j.jcis.2017.04.013.

Aerogels II: Preparation, Properties and Applications
Materials Research Foundations **97** (2021) 57-76

Materials Research Forum LLC
https://doi.org/10.21741/9781644901298-4

Chapter 4

Aerogels for Insulation Applications

M. Ramesh[*1], L. Rajeshkumar[2], D. Balaji[2]

[1]Department of Mechanical Engineering, KIT-Kalaignarkarunanidhi Institute of Technology, Coimbatore-641402, Tamil Nadu, India

[2]Department of Mechanical Engineering, KPR Institute of Engineering and Technology, Coimbatore-641407, Tamil Nadu, India

[*] mramesh97@gmail.com

Abstract

Aerogels have been used as a heat insulating material for the last few decades and possess extremely remarkable qualities for heat insulation. The qualities like light weight (contains more air) and easy to blend with other materials make the aerogel a better insulator than any other material of this kind. The aerogel provides 2 to 3 times better insulation than the styrofoam which is also light weight. The insulating property gets enhanced, if the aerogel is a composite. The interesting details about the insulation property of aerogel will be explored in this chapter.

Keywords

Aerogels, Insulation, Thermal Insulator, Insulation Materials, Eco-Friendly, Sustainable Development

Contents

1.   Introduction

Agricultural waste that causes hazardous effects in the vicinity is considered and the enormous by-products that turn the waste into better-engineered functional results are seen as a sustainable solution in the developing countries [1]. There is a growing interest in the development of innovative heat insulation materials, because traditional insulation materials can not handle heat intelligently [2]. The aerogel is the theoretical product of a work by Kistler (1931). Chalmers [3] describes that Kistler's experiment that the contraction was caused by capillary action during the drying process leads to the formation of a gel. The gel evaluation is performed by heating the sample gels to the critical temperature range along with maintaining the pressure above the liquid vapor level. While raising the temperature above the critical level, converting the liquid to the state of the gases would affect the gel 's structure, it simply dries off inside the whole structure. So, it maintains the gel form. This experiment is replicated over different variants of the gels and has been found positive.

Perrut et al. [4] describes that the aerogels have an extensive diversity of excellent physical nature because of their uncommon morphologies. They are tremendously porous, larger surface area materials, developed from inorganic or organic gels by super critical drying. Aerogels which are based on polymer (organic) are primarily constructed by formaldehyde with resorcinol or melamine, polyisocyanurates, polyurethanes, etc. Carbon aerogels result from particular aerogels of organic nature which are obtained by pyrolysing in the inert environment. The inorganic aerogels are primarily manufactured using the oxides of metal or combination of oxides of metal and materials based on silica. It is recognized that they are remarkable noise control agents, since aerogel can absorb the maximum noise from sound (in the form of pressure energy) through thermal energy convention [5]. In many cases, aerogels were able to reduce the speed of the sound from 340 m/s to 100 m/s, suggesting their strong acoustic insulation [5, 6]. The aerogel has been evolved widely and got so many appellations like worlds lightest solids, solid smoke, air filed solid (70-95%) and frozen smoke. While dropping these materials sounds like metal and one area to improve is brittleness. These issues resolved effectively by making it a composite material by hybridizing with different materials like fiber glass, organic polymers like cellulose extracted from algae, agar and others, PI/SiO$_2$, silica, carbon, metal oxide, cadmium selenide quantum dots and fibrous batting. This hybridizing enhances the overall capability of aerogel in terms of thermal insulation and takes different classification based on many features which are shown in the Fig. 1 [7]. Aerogels can be classified on the based on others parameters are shown in Fig. 2 [8].

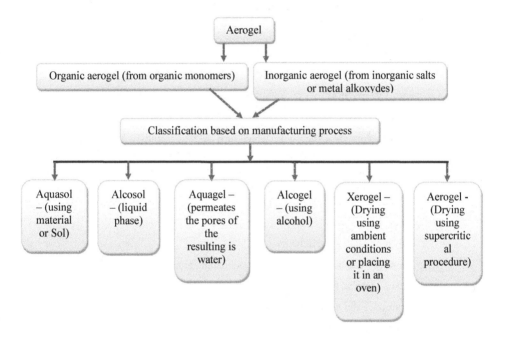

Fig. 1 *Classification of aerogel based on manufacturing process [7].*

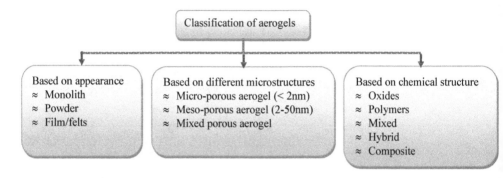

Fig. 2 *Classification of aerogel based other attributes [8].*

The aerogels are composite aerogels which are taking lucrative advantages which occupies many industries nowadays. Noticeably the construction industry, aerospace and aviation industries and coating industry are remarkable. Some of the advantages are list below to make more curious to learn about aerogel. The list is not restricted only to this:

➢ Insulation for building

➢ Chemical absorber

➢ Catalyst or catalyst carrier

➢ Silica based can be used in light guides imaging and optics

➢ Filtering material to remove metals (even heavy)

➢ In cosmetics and paints, it is used to thickening

➢ Machining the impedance

➢ Thermal blankets

➢ Space dust trap

➢ Undergarments

➢ Detector

➢ Manufacture precursor

➢ Electromagnetic shielding

➢ Drug industry as delivering unit to make it as biocompatible

➢ Superconductor constitute

➢ Sports rackets

➢ Water purifying agent

➢ Transmission tunnel as thermal insulator

This chapter only deals with the aerogel insulation applications. The successful insulation requires the material to possess unique properties. The thermal insulation is taken to determine this property has core aerogel insulation. So, this chapter discusses the thermal insulation property of the aerogel. Phalippou et al. [9] states that the aerogel possess extremely high surface area which is 1000 square meter per gram, which made the material porous along with that transparent so it acts as a remarkable material for insulation of heat.

2.   Aerogel as insulating material

An extremely porous yet durable arrangement of aerogels made up of cellulose can be produced using a suitable adhesive agent, with respectable mechanical strength and extreme absorption ability. Aerogels are a rare material class with more than 90 percent porosity. Through them, they virtually eliminated the two major methods of heat transfer that is convection and conduction [1]. According to a study from Grand View Research [10] one of the sections currently engaged by the aerogel industry is the thermal insulation business. A novel innovative application of aerogels based on cellulose from agricultural residues is developed by combining the good thermal insulation competence of cellulose aerogels and their prospective market share. Anisotropic materials through an aligned porous arrangement unveil good insulation property than the isotropic materials, as an isotropic thermal insulating material commonly agonize from localized heat which made inappropriate for managing the thermal energy. Aerogels of anisotropic in nature will significantly reduce the heat transfer and show tremendous insulation in a specific orientation, although some heat may disperse in alternative direction to escape from the confined heat concentration, which may upsurge the overall barrier of heat transfer, the consequence is that it becomes better thermal insulation compared to the isotropic materials [11]. The anisotropic assembly can be erected either through a bottom-up or top-down method. As regards the approach by top-down, the most copious materials in nature is wood, provides attractive medium for the manufacture of anisotropic materials and a fascinating thermal property can be attained by controlling anisotropic wood edifice [12].

Aerogels II: Preparation, Properties and Applications   Materials Research Forum LLC
Materials Research Foundations 97 (2021) 57-76   https://doi.org/10.21741/9781644901298-4

3.   Processing of insulation aerogel

The considerable key segment of developing countries is agriculture. The growing demand for food production, along with a high birth rate and population, results in large quantities of waste as contaminants. Such contaminants lead to numerous environmental glitches as they are destroyed by burning, left for decomposition or directly land-filled [13]. Fluidized bed approach provides efficient transfer of heat and mass between the atomized fluid-gas stream and layer of solid, creating an unvarying coating with nominal cluster. The parameters that regulate the coating consistency of the fluidized bed are the coating solution's surface wettability, viscosity, aerogel particle size and liquid–solid contact time [14]. In this method initially, almost spherical aerogel particles were synthesized of diameter varies between 0.5 and 3 mm using sodium silicate as silica pre-cursor through a rapid sol-gel method. The exterior of the aerogel particles was subsequently coated physically with poly-vinyl alcohol (PVA) which is of aqueous in nature, by using fluidized bed coating method at the bottom of the spray [15]. The schematic arrangement of fluidized coating setup is revealed in Fig. 3 [15]. The fluidized bed system comprises of a cylindrical bottom plate which is of 10 cm in diameter and 5 cm in height wherein the bottom spray nozzle, air distribution plate, and two horizontal gas/suspension inlets is mounted. This is linked to the conical fluidization compartment with the column (Wurster) inside the fluidization chamber.

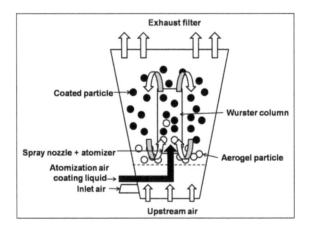

Fig. 3 Scheme of Wurster fluidized bed coater [15].

Cellulose aerogels which are flexible in nature made from originated pineapple-leaf fibers are successfully built using the PVA, as a solvent deionized water is used and adhesive agent followed by freezing [1]. The thermal insulated sheath is constructed based on the sandwich assembly that encapsulates aerogel in between the nylon layer and neoprene outer layer. The nylon material is chosen amongst the fabrics due to its low thermal conductivity, low cost, smooth texture and high availability. Neoprene is also chosen for the waterproof, extreme resistance to tears and its lower thermal conductivity [16].

4.   Thermal insulation properties

Thermal insulation plays a pivotal role in boosting energy efficacy and reducing worldwide energy consumption [17]. For numerous construction and aerospace applications, wherein the heat transfer should be strictly prohibited [18], the production of materials with greater thermal insulation efficiency is extremely desirable. A material acts as a thermal insulator that material should have low thermal conductivity, as is the known fact for solid materials once the material density increases, which increases the thermal conductivity relatively. The mandate here is to diminish the thermal conductivity of the material to act as better insulator. In aerogel concern, the considerable thing is that it has more air in the solid which has poor thermal conductivity than the solid material. The case here is to maintain the air inside constantly while varying the temperature or when any other impact happens to it. The crucial cases replace the air with vacuum or other gas or by sinking the size of the pores underneath the mean free path of air. The maintenance of the state is not so easy, the better alternative is that of nano pores aerogels. The bio-polymer based materials shows the relatively best thermal conductivity value, which could be revealed by the tabulated values in Table 1 of different aerogels. The following table provides the value of thermal conductivity (λ) of different aerogels. The density versus thermal conductivity for polyurethane aerogel is presented in Fig. 4 [27].

Table 1 *Thermal conductivity of numerous materials as aerogel constituent.*

S. No	Name of the material	Thermal conductivity (λ in mW/m/K)	Ref.
1	Fossil-fuel-based foams	40–50	[19]
2	Glass fiber	33–44	[20, 21]
3	Expanded polystyrene	30–40	[20, 21]
4	Mineral wool	30–40	[20, 21]
5	Air	25	[20, 21]
6	Polyurethane	20–30	[20, 21]
7	Fibrous aerogels	18	[22]
8	Silica aerogels	17–21	[20, 21]
9	Aramid aerogels	0.028	[23]
10	Styrofoam-like	0.031	[24]
11	Pectin aerogels	0.016 to 0.022	[25]
12	Cellulose aerogels	0.016	[26]

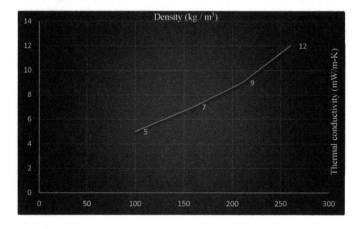

Fig. 4 *Thermal conductivity versus density for polyurethane aerogel [27].*

4.1   Insulation of aerogel as composite material

4.1.1   Solid phase

Aerogels, one of the traditional thermal insulation ingredients, have recently attracted considerable attention [28]. Solid aerogels are derived from gels wherein the solvent is replaced by air, so that stable nano-structures remain intact [29]. The thermal conductivity property is swayed by the volume fraction percentage of aerogel. The volume fraction increase which tends to decrease the thermal conductivity. As known fact, lessening the thermal conductivity increase the insulating ability of the material. Schwertfeger et al. [30] experimented the addition of thermoplastic polyvinylbutyrale (PVB) as binding solid agent owing to its excellent thermal conductivity property. Mixing up of these particles with aerogel and applying pressure up to the melting range of the thermoplastic polymer, which forms a plate. The fig. 5 [30] revels the lowering of volume fraction lowers the thermal conductivity. This could depicts that the influencing nature of the volume fraction over the thermal conductivity of materials.

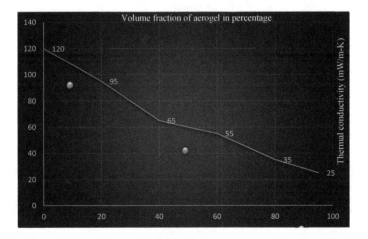

Fig. 5 *PVB plate thermal conductivity vs. volume fraction [30].*

To make aerogels with higher thermal stability, the addition of a material can be both organic and inorganic. Zhou et al. [31] describes the poor mechanical strength of the materials which make the hybrid composites with inorganic materials such as silica. This issue is addressed effectively by the organic hybrid materials which are improved both mechanically and thermally, but the shape is not easy to achieve. One such a kind of experiment is conducted with nano-fibers extracted from bio-cellulose is effectively embedded to form metal-organic framework (MOFs). This composition is very hard to make, but which are remarkable characteristic to confront against the fire. In this study the step by step process of making the hybrid aerogel has been explained. The major properties of MOF are presented in Fig. 6 [31]. The following materials are required to manufacture the MOFs. i) Cladophora cellulose powder, ii) Aluminum nitrite nonahydrate (Al $(NO_3)_3 \cdot 9H_2O$), iii) Terephthalic acid, iv) sodium hydroxide (NaOH) and v) polyvinylpyrrolidone (Mw = 3.6 × 104 g mol^{-1}). Zhou et al. [32] stated that about the preparation of CNF@Al-MIL-53 nano-fibers are illustrated in Fig. 7 [32]. Zhou et al. [31] reported that about the manufacturing of CNF@Al-MIL-53 (CAM) aerogel are illustrated in Fig. 8 [31].

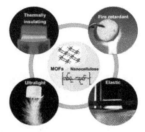

Fig. 6 Property of MOF [31]

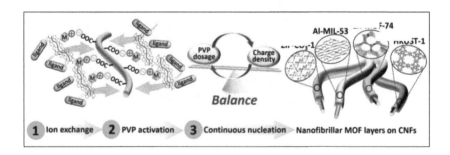

Fig. 7 Representation of the synthesis techniques for the hybrid nanofibers of CNF@Al-MIL-53S [32].

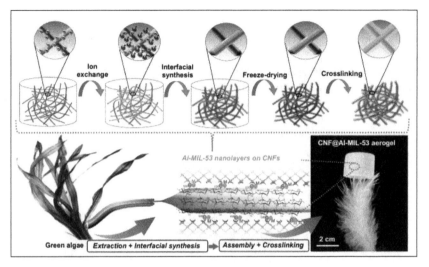

Fig. 8 *Schematic illustration procedures for CAM aerogel [31].*

The above method of making CAM aerogel underwent different stages in the cross-linking process, overcoming the fragile existence of the inorganic aerogel. It increases the mechanical strength with good compression modulus of approximately 200 MPa cm^{3}g^{-1}, and specific stress of approximately 100 MPa cm^{3}g^{-1}. The experiment reveals that the aerogel which is lighter weight with effective cellular framework structure also possess ordered porosity. The other considerable parameters are low thermal conductivity, flame resistant and moisture resistant. Another considerable work in the solid phase is carried out by Fan et al. [33] in developing the composite PI/SiO$_2$ aerogels and experimenting it. The manufacturing process is illustrated in Fig. 9 [33].

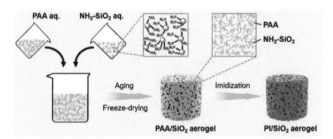

Fig. 9 *Schematic illustration for manufacturing the PI/SiO$_2$ aerogel [33].*

The capability of the PI/SiO$_2$ composite aerogels for the dimensional, mechanical and thermal stability have been assessed. The images presented in Figs. 10 & 11 reveal that these could prevent shrinkage less than 20% and maintained high porosity (more than 95%) of aerogel, resultant in PI/SiO$_2$ hybrid composite possess good dimensional stability and an even porous construction. The mechanical property is concern possess better stress strain range along with the specific modulus more than 100 KNmkg^{-1}. It terms of thermal property is concern it possess considerable low thermal conductivity along with high operating temperature of around 450 °C.

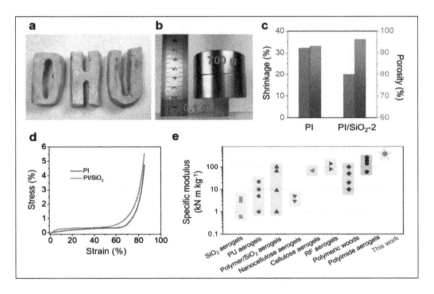

Fig. 10 *Mechanical and dimensional characteristic of PI/SiO$_2$ aerogel [33].*

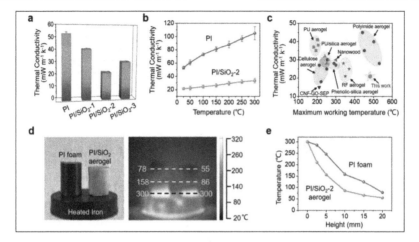

Fig. 11 Performance of thermal insulation of PI/SiO₂ aerogel [33].

4.1.2 Liquid phase

Schwertfeger et al. [30], experimented the vinyl acetaterethe (VA) liquid is dispersed with the aerogel. This mixture is then subjected to pressure to form the plate. Plate produced is allowed to dry, and the dispersed particles produce a polymer film that binds to the aerogel during drying. The measurement is taken for the thermal conductivity with the volume fraction, the following graph is drawn.

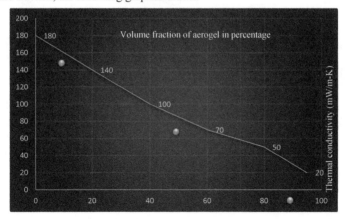

Fig. 12 Liquid VA plate thermal conductivity vs. volume fraction [30].

5.   Applications of insulation aerogels

The thermal insulation nature of the aerogel is considerably remarkable, which results in different applications of aerogels. The image presented in Fig. 13 [34-36] describes the various applications of aerogel specifically for thermal insulation applications. The thermal insulation is necessary wherever the process of burning occurs, which is high for the aviation and aerospace. The insulation requirement is over and above 2000 °C too.

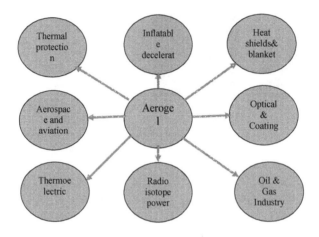

Fig. 13 *Various application of thermal insulating aerogel [34-36].*

Human always urge towards moving in high speed. Any vehicle which could travel at high speed requires fuel. While burning it liberates enormous heat, this heat can be controlled only by the insulation. This industry is very keen in reducing the weight. So, the only light weight effective thermal insulator is none other than the aerogel. Table 2 [37-40] depicts the various types of aerogel used for various locations in aviation and aerospace industry. The aerogels shown in the table 2 for aerospace and aviation application are mostly used as blankets. The images presented in Figs. 14 [41] and 15 [42] are the examples for aerogel based blankets.

Table 2 *Application of aerogel in aviation and aerospace*

S. No	Name of the material	Application	References
1	Silica aerogel blanket	In compartments as thermal barrier	[37]
2	Silica aerogel blanket	Fire retardation	[37]
3	Silica aerogel	Hypervelocity Particle Capture	[37]
4	Silica aerogel	Thermal insulation for Mars Rover	[37]
5	Silica aerogel	Cryogenic fluid containment	[37]
6	Aluminum and Silica aerogel	Laminate thermal insulation blanket for aircraft	[38]
7	Silica aerogel	Inflatable decelerator concept for EDL	[39]
8	Silica aerogel	Extra-Vehicular Activity (EVA) suit	[39]
9	Polymer cross-linked aerogels	Enhanced mechanicalproperties for aerospace	[40]
10	Flex-link aerogels	Flexibility and durability to aerospace materials	[40]
11	Reinforced aerogels	Low thermal conductivity	[40]

Fig. 14 *Aerogel blanket SEM image [41].*

The major highlights of this blanket are: i) hybrid aerogel blankets can be synthesis in single go, so process of Resorcinol-formaldehyde-APTES is easier, ii) shrinkage of aerogel is limited owing to the usage of polyethylene terephtalate mat as core for blanket, and iii) In room conditions, the thermal conductivity is low around 0.019 $Wm^{-1}K^{-1}$ obtained.

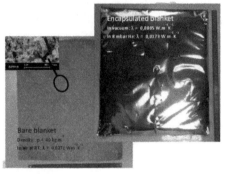

Fig. 15 *Aerogel blanket [42].*

The major highlights of this blanket presented in Fig. 15 are: i) space applications required flexible and lightweight material which satisfies the need, ii) at room temperature, density = 0.04 g cm^{-3} and $\lambda = 0.027$ $Wm^{-1}K^{-1}$ for flexible hybrid blanket, iii) at vacuum, encapsulated blanket got λ value of 0.0005 $Wm^{-1}K^{-1}$, and iv) An encapsulated blanket at 8 mbar He, has got the λ value of 0.0173 $Wm^{-1}K^{-1}$.

Conclusions

In this chapter, an innovative class of anisotropic material, lightweight and mechanically robust aerogel with outstanding thermal insulation capability has been discussed. From the literature it is found that the core–shell structure thermal conductivity was lower than that of the aerogel owing to better thermal insulation and thin PVA coating. The thermal components heat-insulation efficiency is about 3 times better than that of a commercially existing product. The unwanted sound reduction coefficient of aerogel samples with the same thickness is superior to that of an existing Basmel absorber which exhibits its potential in sound insulation applications. The results show great potential in acoustic and heat insulation applications for the aerogel. From the chapter it is concluded that the cellulose-based aerogels are showed to have a low thermal conductivity. In many ways the aerogels can be make as a better insulating materials, the key lies here is manufacturing method, that could be clearly understood by the methods explained. Amongst of all other performance this study focuses on the thermal insulation, there is no second thought that the aerogel shows its superiority while making it has hybrid composite material.

References

[1] N.H.N. Do, T.P. Luu, Q.B. Thai, D.K. Le, N.D.Q. Chau, S.T. Nguyen, P.K. Le, N. Phan-Thien, H.M. Duong, Heat and sound insulation applications of pineapple aerogels from pineapple waste, Mater. Chem. Phy. (2019), https://doi.org/10.1016/j.matchemphys.2019.122267

[2] X. Zhang, X. Zhao, T. Xue, F. Yang, W. Fan, T. Liu, Bidirectional anisotropic polyimide/bacterial cellulose aerogels by freeze-drying for super-thermal insulation, Chem. Eng. J. (2019). https://doi.org/10.1016/j.cej.2019.123963

[3] Chalmers, Civil and Environmental Engineering, Report 2012:2.

[4] M. Perrut, Eric françaiseparex, Aerogel drying, 5, rue Jacques Monod F-54250 Champigneulles.

[5] S. Malakooti, H.G. Churu, A. Lee, T. Xu, H. Luo, N. Xiang, C. Sotiriou-Leventis, N. Leventis, H. Lu. Sound insulation properties in low-density, mechanically strong and ductile nanoporous polyurea aerogels, J. Non-Cryst. Solids. 476 (2017) 36-45. https://doi.org/10.1016/j.jnoncrysol.2017.09.005

[6] Y. Lu, X. Li, X. Yin, H.D. Utomo, N.F. Tao, H. Huang. Silica aerogel as super thermal and acoustic insulation materials, J. Environ. Prot. 9 (2018) 295-308. https://doi.org/10.4236/jep.2018.94020

[7] R.J. Ayern, P.A. Iacobucci, Metal oxide aerogel preparation by supercritical extraction, Rev. Chem. Eng. 5 (1988) 157-198. https://doi.org/10.1515/REVCE.1988.5.1-4.157

[8] A. Du, B. Zhou, Z. Zhang, J. Shen. A special material or a new state of matter: are view and reconsideration of the aerogel, Mater. 6(3) (2013) 941–968. https://doi.org/10.3390/ma6030941

[9] J. Phalippou, R. Vacher, Aerogels Proceedings of the Fifth International Symposium on Aerogels (ISA-5) Montpellier, France. 8-10 September 1997.

[10] Grand View Research. Aerogel Market Size, Share & Trends Analysis Report By Form (Blanket, Particle, Panel, Monolith), By Product (Silica, Carbon, Polymers), By End Use, By Technology, By Region, And Segment Forecasts, 2018 - 2025. 2018 July 11, 2019]; Available from: https://www.grandviewresearch.com/industry-analysis/aerogelmarket.

[11] T. Li, J. W. Song, X.P. Zhao, Z. Yang, G. Pastel, S.M. Xu, C. Jia, J.Q. Dai, C.J. Chen, A. Gong, F. Jiang, Y. G. Yao, T.Z. Fan, B. Yang, L. Wågberg, R.G. Yang, L.B. Hu. Anisotropic, lightweight, strong, and super thermally insulating nanowood with naturally aligned nanocellulose, Sci. Adv. 4 (2018) 3724-3733. https://doi.org/10.1126/sciadv.aar3724

[12] T. Li, Y. Zhai, S.M. He, W. T. Gan, Z.Y. Wei, M. Heidarinejad, D. Dalgo, R.Y. Mi, X.P. Zhao, J.W. Song, J.Q. Dai, C.J. Chen, A. Aili, A. Vellore, A. Martini, R.G. Yang, J. Srebric, X.B. Yin, L. B. Hu. A radiative cooling structural material, Science 364 (2019) 760-763. https://doi.org/10.1126/science.aau9101

[13] N.A. Ruslan, N.F.M. Aris, N. Othman, A.R. Saili, M.Z. Muhamad, N.N.H. Aziz. A Preliminary study on sustainable management of pineapple waste: Perspective of smallholders, Int. J. Acad. Res. Bus. Soc. Sci. 7(6) (2017) 1-7. https://doi.org/10.6007/IJARBSS/v7-i6/2937

[14] B. Guignon, A. Duquenoy, E.D. Dumoulin. Fluid bed encapsulation of particles: Principles and practice, Dry. Technol. 20(2) (2002) 419–447. https://doi.org/10.1081/DRT-120002550

[15] Z.A.A. Halim, M.A.M. Yajid, M.H. Idris, H. Hamdan. Physiochemical and thermal properties of silica aerogel–poly vinyl alcohol/core–shell structure prepared using fluidized bed coating process for thermal insulation applications, Mater. Chem. Phy. (2018). https://doi.org/10.1016/j.matchemphys.2018.05.019

[16] H.M. Duong, Z.C. Xie, K.H. Wei, N.G. Nian, K. Tan, H.J. Lim, A.H. Li, K.S. Chung, W.Z. Lim. Thermal jacket design using cellulose aerogels for heat insulation application of water bottles, Fluids 2(4) (2017) 64. https://doi.org/10.3390/fluids2040064

[17] X. Xu, Q.Q. Zhang, M.L. Hao, Y. Hu, Z.Y. Lin, L.L. Peng, T. Wang, X.X. Ren, C. Wang, Z.P. Zhao, C.Z. Wan, H.L. Fei, L. Wang, J. Zhu, H.T. Sun, W.L. Chen, T. Du, B. W. Deng, G.J. Cheng, I. Shakir, C. Dames, T.S. Fisher, X. Zhang, H. Li, Y. Huang, X.F. Duan, Double-negative-index ceramic aerogels for thermal superinsulation, Science 363 (2019) 723-727. https://doi.org/10.1126/science.aav7304

[18] J.Y. Zhang, Y.H. Cheng, M. Tebyetekerwa, S. Meng, M.F. Zhu, Y.F. Lu. Stiff–soft binary synergistic aerogels with super flexibility and high thermal insulation performance, Adv. Funct. Mater. 29 (2019) 6407-6418. https://doi.org/10.1002/adfm.201806407

[19] W.S. Chen, Q. Li, Y.C. Wang, X. Yi, J. Zeng, H.P. Yu, Y.X. Liu, J. Li. Comparative study of aerogels obtained from differently prepared nanocellulose fibers. ChemSusChem 7 (2014) 154–161. https://doi.org/10.1002/cssc.201300950

[20] B.P. Jelle. Traditional, state-of-the-art and future thermal building insulation materials and solutions—Properties, requirements and possibilities. Energy Build. 43 (2011) 2549–2563. https://doi.org/10.1016/j.enbuild.2011.05.015

[21] J.E. Fernandez. Materials for aesthetic, energy-efficient, and self-diagnostic buildings, Science 315 (2007) 1807–1810. https://doi.org/10.1126/science.1137542

[22] D. Bendahou, A. Bendahou, B. Seantier, Y. Grohens, H. Kaddami, H. Nano-fibrillated cellulose-zeolites based new hybrid composites aerogels with super thermal insulating properties, Ind. Crop. Prod. 65 (2015) 374–382. https://doi.org/10.1016/j.indcrop.2014.11.012

[23] N. Leventis, C. Chidambareswarapattar, D.P. Mohite, Z.J. Larimore, H.B. Lu, C. Sotiriou-Leventis. Multifunctional porous aramids (aerogels) by efficient reaction of carboxylic acids and isocyanates, J. Mater. Chem. 21 (2011) 11981–11986. https://doi.org/10.1039/c1jm11472g

[24] N. Leventis, C. Sotiriou-Leventis, D.P. Mohite, Z.J. Larimore, J.T. Mang, G. Churu, H.B. Lu. Polyimide aerogels by ring-opening metathesis polymerization (ROMP), Chem. Mater. 23 (2011) 2250–2261. https://doi.org/10.1021/cm200323e

[25] B. Wicklein, A. Kocjan, G. Salazar-Alvarez, F. Carosio, G. Camino, M. Antonietti, L. Bergstrom. Thermally insulating and fire-retardant lightweight anisotropic foams based on nanocellulose and graphene oxide, Nat. Nanotechnol. 10 (2014) 277–283. https://doi.org/10.1038/nnano.2014.248

[26] C. Rudaz, R. Courson, L. Bonnet, S. Calas-Etienne, H. Sallee, T. Budtova. Aeropectin: Fully biomass-based mechanically strong and thermal super insulating aerogel. Biomacromol. 15 (2014) 2188–2195. https://doi.org/10.1021/bm500345u

[27] G. Biesman, D. Randall, E. Francais, M. Perrut. Polyurethane-based organic aerogels' thermal performance, J. Non Crystal. Solids 225 (1998) 36-40. https://doi.org/10.1016/S0022-3093(98)00103-3

[28] R.L. Liu, X. Dong, S.T. Xie, T. Jia, Y.J. Xue, J.C. Liu, W. Jing, A.R. Guo, Ultralight, thermal insulating, and high-temperature-resistant mullite-based nanofibrous aerogels, Chem. Eng. J 360 (2019) 464-472. https://doi.org/10.1016/j.cej.2018.12.018

[29] L.Z. Zuo, W. Fan, Y.F. Zhang, L.S. Zhang, W. Gao, Y.P. Huang, T.X. Liu, Graphene/montmorillonite hybrid synergistically reinforced polyimide composite aerogels with enhanced flame-retardant performance, Compos. Sci. Technol. 139 (2017) 57-63. https://doi.org/10.1016/j.compscitech.2016.12.008

Materials Research Forum LLC
https://doi.org/10.21741/9781644901298-4

[30]   F. Schwertfeger, D. Frank, M. Schmidt, Symposium on Aerogels, 8–10th September 1997, Montpellier, France.

[31]   S. Zhou, V. Apostolopoulou-Kalkavoura, M.V. da Costa, L. Bergström, M. Strømme, C. Xu. Elastic aerogels of cellulose nano fibers@ metal–organic frameworks for thermal insulation and fire retardancy, Nano-Micro Lett. 12(1) 2020 1-3. https://doi.org/10.1007/s40820-019-0343-4

[32]   S. Zhou, M. Strømme, C. Xu. Highly transparent, flexible and mechanically strong nanopapers of cellulose nanofibers@metal–organic frameworks, Chem. Eur. J. 25 (2019) 3515–3520. https://doi.org/10.1002/chem.201806417

[33]   W. Fan, X. Zhang, Y. Zhang, Y. Zhang, T. Liu. Lightweight, strong, and super-thermal insulating polyimide composite aerogels under high temperature, Compos. Sci. Technol. 173 (2019) 47-52. https://doi.org/10.1016/j.compscitech.2019.01.025

[34]   https://technology.grc.nasa.gov/patent/TOP3-413.

[35]   U.S. Department of Energy, Guiding principles for sustainable federal buildings; https:// energy.gov/eere/femp/guiding-principles-sustainable-federal-buildings.

[36]   M. S. Al-Homoud, Performance characteristics and practical applications of common building thermal insulation materials, Build. Environ. 40 (2005) 353–366. https://doi.org/10.1016/j.buildenv.2004.05.013

[37]   N. Bheekhun, A. Talib, A. Rahim, M.R. Hassan. Aerogels in aerospace: an overview, Adv. Mater. Sci. Eng. 2013. https://doi.org/10.1155/2013/406065

[38]   US patent application - US20120308369A1 with tilted "Laminate thermal insulation blanket for aircraft applications and process there for" in 2012.

[39]   J.P. Randall, M.A. Meador, S.C. Jana. Tailoring mechanical properties of aerogels for aerospace applications, ACS Appl. Mater. Interf. 3(3) (2011) 613-626. https://doi.org/10.1021/am200007n

[40]   https://ntrs.nasa.gov/archive/nasa/casi.ntrs.nasa.gov/20080047425.pdf (Michael A. Meador-2008)

[41]   S. Berthon-Fabry, C. Hildenbrand, P. Ilbizian. Lightweight super insulating resorcinol-formaldehyde-APTES benzoxazine aerogel blankets for space applications, Euro. Polym. J. 78 (2016) 25-37. https://doi.org/10.1016/j.eurpolymj.2016.02.019

[42]   S. Berthon-Fabry, C. Hildenbrand, P. Ilbizian, E. Jones, S. Tavera. Evaluation of lightweight and flexible insulating aerogel blankets based on resorcinol-formaldehyde-silica for space applications, Euro. Polym. J. 93 (2017) 403-416. https://doi.org/10.1016/j.eurpolymj.2017.06.009

Aerogels II: Preparation, Properties and Applications
Materials Research Foundations97 (2021) 77-98

Materials Research Forum LLC
https://doi.org/10.21741/9781644901298-5

Chapter 5

Aerogels as Catalyst Support for Fuel Cells

R. Imran Jafri[*], Soju Joseph, Akshaya S Nair

Department of Physics and Electronics, CHRIST (Deemed to be University), Hosur Road, Bengaluru-560029, India

* imran.jafri@christuniversity.in

Abstract

Environmental pollution caused by the extensive use of fossil fuels and global energy crisis have increased the need to look for renewable energy sources that not only supplement the global energy needs but are economical and environment friendly, thus making way for fuel cells (FCs) as one of the alternatives for replacing the existing fossil fuel based machinery. Nevertheless, there are several factors that account for the hindrance of FCs on a large scale, one of them being the sluggish oxygen reduction reaction (ORR) kinetics taking place at the cathode. Aerogels are a class of promising materials that have the potential to improve the electrocatalytic activity, stability and durability of FCs when used as catalyst support. The present chapter focuses on reporting the latest developments in the field of aerogels as catalyst support for FCs.

Keywords

Oxygen Reduction Reaction, Fuel Cells, Aerogels, Electrocatalytic Activity, Durability, Stability

Contents

1.   Introduction

The need for developing energy storage device with the rise in the global energy demand has prompted researchers to explore fuel cell (FC) which converts chemical energy of the fuel to electricity and resulting in environment friendly products. Their high efficiency and facile nature have attracted huge attention world-wide [1-3]. Among the different types of FCs, polymer electrolyte membrane (PEM) FCs have the potential to eradicate the growing energy and environment related issues, but researchers are facing obstacles in establishing FCs that are profitable and simultaneously economical. The three main obstacles that are hindering the FC commercialization on a larger scale are: cost, inadequate activity, and insufficient desirable durability [4]. The durability of FCs depends on the ORR that takes place at the cathode. The key to boost the FC commercialization lies in developing active electrocatalyst that can not only improve the sluggish ORR kinetics at the cathode but also reduces the total cost of FC system. Platinum (Pt) based catalyst exhibit high durability and activity towards ORR, but their high cost and scarcity overshadows the performance [5]. Intensive research is put forth to replace the Pt based catalyst with other carbon-based materials, metal oxides, metal aerogels, transition metals, heteroatom doped atoms as catalyst support that can meet the above- mentioned requirements. After the initial reports by Kistler in 1930s on aerogels [6], several other forms of materials like oxide aerogels [7], chalcogenide aerogels [8], and carbon-based aerogels [9-12] have been exploited for different applications ranging from energy storage devices to thermal insulators [11].

Extensive research has also been done on various support-less catalyst such as nanowires, nanotubes, metal aerogels (MA) etc., for enhancing the durability and catalytic activity [13]. Their peculiar properties include high-surface area, porous nature, robustness and exclusion from support corrosion which is the major reason for instability in FC devices. Meanwhile, MA consist of an extended metal backbone nano network, and possess both the exceptional properties of a metal, like high electrical and thermal conductivity followed by the properties of aerogels (high-surface area, porosity and ultralow density) which makes them a potential candidate to be utilized as a catalyst for improving the performance of FCs [14-16].

With the advancement in FC technology, many types of aerogels are being studied based on various properties that are unique to them. The structures are also being tailored alongside to produce 3D mesoporous entities [17] possessing a high density of interconnected porous networks along with doping and co-doping to increase the catalytically active regions. Doping helps in the attachment of catalyst particles over the support preventing agglomeration during consecutive working cycles. Doping also imparts additional catalytic activity without affecting the physicochemical properties of the pristine graphene [18].

Graphene has stood up as a great contender for the development of graphene aerogels (GAs) as catalyst support showing great promise by showing a dramatic increase in the current and power densities [19]. However, it also possesses a problem of restacking with prolonged usage. This study also discusses various developments in the technology to prevent restacking and increase durability. The 3D GAs have also shown to reduce the water flooding experienced by FCs and prevent fuel crossover and thereby showing tolerance to poisoning.

There are various FC organization that sets technical target and test protocols for each of the FC components. US DOE (Department of Energy) being one of them have set technical target for electrocatalytic activity of FC. The target for 2025 is as follows: Developing electrocatalysts with reduced PGM (Pt group metals) loading of ≤ 0.10 g/kW and it should be durable for 8000 hours of operation [20].

2.   Carbon based aerogels

2.1   Carbon aerogels

Carbon aerogels (CAs) are synthesized using supercritical drying and pyrolysis of an organic gel. One pot method was utilized to synthesis WC@C/N/CA-doped porous catalyst by Hong Zhu et al. [21]. Tungsten was effectively topped into CAs and shaped a core shell structure (WC@C) treated by calcination. WC@C/N/CA-doped catalyst showed better selectivity, stability, large surface area and activity as it exhibits nearly 4e$^-$ ORR process in alkaline media. Doping or stacking WC could improve the active site density of catalyst. CAs have large mesoporous volumes and high surface area and thus are of interest for energy storage applications. Fig. 1 [22] shows the schematic setup of a direct methanol FC or DMFC.

Materials Research Forum LLC
https://doi.org/10.21741/9781644901298-5

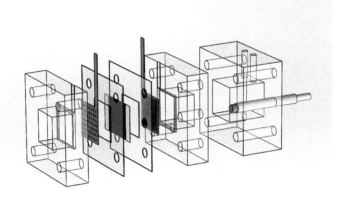

Fig. 1 Schematic of experimental setup for DMFC. [Reprinted with permission from Ref. [22] Copyright (2019) American Chemical Society].

A microwave-assisted polyol process was used by Hong Zhu et al. [23] to synthesize Pt-Au based on carbon aerogel (CA). PtAu/CA showed an I_f/I_b ratio higher than its competitor indicating efficient removal of CO species, where I_f and I_b represent the forward and backward peak current densities during methanol oxidation respectively, higher ratio indicates complete oxidation of methanol to CO_2. The mesoporous structure facilitates mass transfer and proton diffusion along with a high methanol tolerance and catalytic activity.

Bong S. et al. [24] pointed out that layer thickness and catalyst loading are pivotal in governing the performance of the FC and carbon-based aerogels are competing candidates possessing a large surface area for optimal dispersion of catalyst over it [24]. They developed a stable electrocatalyst via poly-condensation of resorcinol, formaldehyde and Pt and ruthenium catalyst. PtRu/CA having the highest performance among the various electrocatalysts developed possessed an Electrochemical Surface area (ECSA) of 85.8 m^2 g^{-1}. Since, the size of the pore and its volume affect mass transport, sodium carbonate was used to control these factors. Fig. 2 [22] shows water flooding at cathode current collectors, a challenging problem in the working of a DMFC.

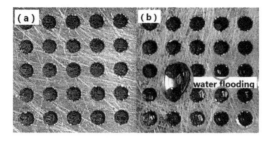

Fig. 2 Typical images of water flooding at surface of cathode current collectors [Reprinted with permission from Ref. [22] Copyright (2019) American Chemical Society].

2.2   Graphene aerogels

Combining metal clusters and organic linkers results in metal-organic frameworks (MOFs) that are crystalline porous materials. Nanostructured materials can be prepared using a facile MOFs-templated synthesis technique. Selective pyrolysis in controlled atmosphere can be used to convert MOFs into porous carbon or metal oxides. But at high temperatures it is difficult to tune the structure of metal-organic framework (MOF) derived materials as the induction force during structure transformation at these temperatures is absent. Wei Xia et al. [25] broke the bulk MOF crystals into monodisperse metal oxide hollow nanoparticles (NPs) (Co based MOF, $Co(mIM)_2$ (mIM=2-methylimidazole)) using a graphene aerogel (GA) assisted technique and is as shown in fig. 3 [25].

WeiXia et al. [25] observed that the original Co-MOF nanocrystals that were obtained from monodisperse cobalt oxide (CoO_x) hollow NPs had a size of 35 nm that were dispersed onto N-doped graphene aerogels (NG-As) and subsequent thermal activation could yield uniform dispersion of these particles on a graphene sheet surface. A new material CoO_x/NG-A possessed excellent electrocatalytic activity for ORR. Also, they attained N rich 3D Carbon material (C/NG-A) which displayed rate capability of 305 F g^{-1} at 50 A g^{-1}, high specific capacitance of 421 F g^{-1} at a current density of 1 A g^{-1}, with an 99.7% capacitance retention after 20,000 cycles of cycling performance, when employed as an electrode material in supercapacitors (SCs).

The idea behind development of various catalyst for DMFC is that the novel catalyst must possess a high ECSA along with high durability and should be cost-effective. YamengWang et al. [26] prepared a 3D aerogel which was composed of graphene and multi-walled carbon nanotubes (MWCNTs). The PtNi based electrocatalyst was prepared via one-step co-reduction of graphene and MWCNTs and showed an ECSA of 17.9 m^2g^{-1} with a mass activity of 35.7 mA mg^{-1}. The kinetic current density was reported as 199.4 μA cm^{-2}. This electrocatalyst showed a higher tolerance towards methanol oxidation reaction (MOR) compared to the standard Pt/C. The formation of an intermetallic compound in the O-Pt-Ni/CA catalyst resulted in negligible change in the mean particle size after an accelerated stress test. The catalyst only suffered a decrement of 29% in its mass activity after 3000 cycles indicating that its 3D structure prevented agglomeration.

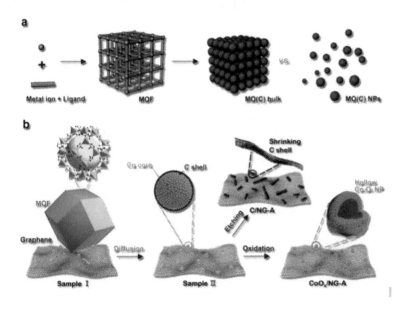

Fig. 3 (a) Design of breaking MOF into monodisperse metal oxide NPs. (b) Schematic representation of the development process of CoOₓ/NG-A and C/NG-A. [Reprinted with permission from Ref. [25]. Copyright (2017) American Chemical Society].

Graphene soon overtook carbon in this field due to its low density which made it super absorbent. It also possesses a very high mechanical strength, superior electrical conductivity and is resistant to thermal changes. Fig. 4 [27] shows images of GAs taken under a high-resolution scanning electron microscope (SEM).

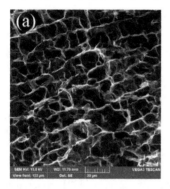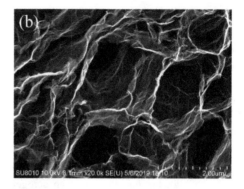

*Fig. 4 SEM images of graphene aerogel [Reprinted with permission from Ref. [27].
Copyright (2019) American Chemical Society].*

A 3D-NGA (nitrogen doped graphene aerogel) possessing high porosity was developed by Shenlong Zhao et al. [28] as support for anode in DMFC with PtRu as a catalyst. The electrocatalyst showed a forward peak current density of 668 mA mg^{-1} with the I_f/I_bratio reaching upto 2.6. The sample exhibited a peak power density of 93 mW cm^{-2}along with large power density which attributed to superior catalytic activity towards MOR. After 3600s, the current density decreased to 43% of the initial which indicated a better performance when compared to other catalysts.

Jialin Duan et al. [29] synthesized Pt based graphene oxide aerogel (GOA) possessing a 3Dmacroporous structure which provided a larger surface area for MOR and showed an accelerated mass transfer. The ECSA of Pt/GOA electrocatalyst was 95.5 m^{2}g^{-1} confirming that the support enhanced the catalyst utilization. The peak current density reached 876 mA mg^{-1} Pt greater than its commercial competitors and exhibited better catalytic stability. The use of GOA prevented the stacking of graphene sheets thus improving the catalytic performance for DMFC.

Mingrui Liu et al. [30] synthesized 3D-GAs using a facile hydrothermal method with Pd NPs uniformly distributed over it, which showed greater catalytic activity and superior stability towards MOR. The cyclic voltammetry (CV) of Pd/3DGA showed an ECSA of 425 m^{2}g^{-1} which was 3.4 times better than the commercial Pd/C. The electrocatalyst also showed a forward anodic peak current density of 7.54 A mg^{-1}. After 500 potential cycles, the electrocatalyst showed little change in its morphology. The improved activity of the catalyst was credited to the surface structure, the electrical conductivity due to sp^2 carbon conjugation and high porosity.

A three-dimensional Pt/C/graphene aerogel (Pt/C/GA) hybrid was synthesized by LeiZhao et al. [31] via hydrothermal process which showcased an enhanced stability towards MOR indicating no decrement in the electrocatalytic activity, scavenging the crossover methanol under high potential. The ECSA and mass activity for Pt/C/GA was reported to be a little lower than the commercial Pt/C, as 70.4 m^2 g^{-1} and 405.3 mA mg^{-1} respectively However, it performed far better than the commercial Pt/C, losing only 16% of its initial activity after 1000 CV cycles. This enhancement was attributed to the unique framework of 3D graphene and the efficiency of the assembly between the aerogel and the catalyst.

3D graphene is said to be one of the best catalysts out there due to its superior properties along with high loading volume of an aerogel. Xuan Zhang et al. [32] explored the possibilities of controllable integration of Pt NPs over 3D graphene to synthesize a 3DNGA via one-pot hydrothermal route. This Nitrogen doped 3D structure provided an ECSA of 42.17 m^2 g^{-1}anchoring more active sites for a better MOR performance. The anodic peak current was found to be 9.32 mA cm^{-2} along with a lower onset potential for methanol oxidation implying better electro-catalytic activity and removal of CO intermediates from the catalyst surface. The I_f/I_b ratio was 2.1 indicating better oxidation of fuel compared to other catalysts.

Xinglan Peng et al. [33] developed 3D-NGA based on PtCu alloy as a stable electrocatalyst for DMFC combining hydrothermal method with microwave-assisted polyol process. This catalyst possessed an ECSA of 47.8 m^2 g^{-1} along with a current density that was 4.5 times higher than Pt/XC-72. This enhancement in the performance towards MOR was attributed to the interconnected porous network of the support along with nitrogen doping and strong interaction between the catalyst and the support.

LeiZhao et al. [34] synthesized a 3D nitrogen doped graphene catalyst support using a combination of hydrothermal self-assembly process along with thermal treatment and template-removing. This Pt based electrocatalyst showed a high ECSA of 90.7 m^2 g^{-1} and a better catalytic activity compared to Pt/3D-GA. It also showed a peak current density of 544.5 mA mg^{-1} along with a superior mass activity. After 1000 cycles, the ECSA of Pt/3D-NGA suffered a loss of only 21% retaining its original structure. This indicated less agglomeration and high stability.

A Pt based graphene nanotube composite was developed by Y.H.Kowk et al. [35] with a pore size less than 10μm using a polyol process. The Ru-Pt GO CNT composite had a maximum specific power of 10.15mW mg^{-1} and a maximum power density of 16.35 mWcm^{-2}. The process employed for the production of such an electrocatalyst was environment friendly and thus could be used for mass production.

Huajie Huang et al. [36] developed 3D porous electrocatalyst from gas using a self-assembly method. The 3D $Pt/RuO_2/GA$ electrocatalyst was found to have superior activity, resistance to CO poisoning and higher stability towards MOR. The ECSA of Pt/RuO_2G was found as 122.7 m^2g^{-1} with a mass activity of 841.9 mA mg^{-1} and a stable voltage for about 300s. The continuous pores and large surface area along with uniform dispersion of Pt facilitated electron and proton conductivities.

In 2017, Y.H. Kwok et al. [37] decorated GA with ultra-fine Pt NPs for direct methanol microfluidic FC. This aerogel was annealed in air (to reduce the oxygen functional groups over GO) hereby improving the catalytic performance which resulted in higher ECSA. The ECSA of Pt/GO was found to be 1882 cm^2/mg with a maximum specific power of 6.94 Mw/mg Pt and a conductivity of 28 S/m. The specific power of this electrocatalyst increased by 358% compared to commercial Pt/C. After 20 cycles, there was little decrement in the maximum specific power compared to Pt/C. The porosity of the electrode, further facilitated the conductivity of the electrode achieving better cell performance. Fig. 5 [27] shows the schematics of a test system for methanol crossover.

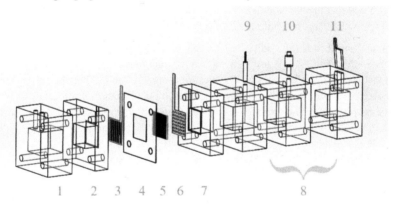

Fig. 5 Schematic diagram of the test system for the methanol crossover of μDMFC. (1) Anode reservoir chamber; (2) anode end plate; (3) anode current collector; (4) gasket; (5) MEA (without cathode GDE); (6) cathode current collector; (7) cathode end plate; (8) cathode chamber; (9) working electrode (GCE); (10) reference electrode; (11) counter electrode. [Reprinted with permission from Ref. [27]. Copyright (2019) American Chemical Society].

A 3D hybrid aerogel was built by Li-Mei Zhang et al. [38] from graphene and polypyrrole-derived N-doped CNTs via hydrothermal process, it was seen that the

Materials Research Forum LLC

https://doi.org/10.21741/9781644901298-5

electrocatalyst showed improvement in the electrocatalytic activity and resulted in higher stability towards MOR than its predecessor 3D GA. The synthesized Pt/G-NCNT showed an ECSA of 118.4 m^2g^{-1} with a higher current density (0.74 A mg^{-1}) than Pt/GA which remained high, even after 3600s of testing. With a higher retention rate, even after 1000 cycles, it was established that the structure of the electrocatalyst increases the active sites for better Pt loading, also providing steric hindrance which prevents its migration. This synergistic effect of GA and N-CNT resulted in the superior stability and electrocatalytic activity for MOR.

YangZhou et al. [39] efficiently freeze-dried slurry of 3D graphene/carbon nanotube (GR-CNTs) with polyvinyl alcohol (PVA) as an organic binder to synthesize 3D GR-CNT aerogels. Using Pt as catalyst, an ESCA of 75.0 m^2g^{-1} was achieved which was much larger than its predecessor Pt/GR; hinting a richness in catalytically active regions. The hybrid exhibited a maximum current density of 550 mA mg^{-1}, approximately 4 times higher than Pt/GR. The ratio of I_f/I_b was found to be highest for Pt/GR-CNTs (1.53) indicating a higher tolerance to accumulation of carbonaceous species during MOR. This study found that the addition of CNT in the aerogel prevented the restacking of graphene sheets providing an interconnected 3D porous structure with a large surface area.

Minmin Yan et al. [40] synthesized carbon nanotube/nitrogen-doped graphene hybrid aerogel (Pt/LDCNT-NG) using a bottom-up approach via self-assembly process demonstrating low defect density. The electronic structure of the electrocatalyst is highly optimized and the Pt stability is enhanced resulting in higher electrocatalytic activity which outperforms the conventional doped carbon and graphene catalysts. The ECSA of Pt/(LDCNT)$_3$-(NG)$_7$ was reported to be 132.4 m^2 g^{-1} with a mass activity of 871.9 mA mg^{-1}. It also showed slower decay rate along with a higher steady-state current density. The catalyst retained 79.1% of its initial mass activity after 100 cycles showing superior tolerance towards MOR.

A state-of-the-art approach was adopted by Xu-LeiSui et al. [41] to synthesize porous N-doped GA comprising of an open structure and having an abundance of defects via hydrothermal self-assembly of GO and zeolitic imidazolate framework (ZIF)-8. An abundance of N-doped sites along with microporous structure is created via introduction of N and Zn. This structure improves the surface area accessible, resulting in high performance of Pt based catalyst towards MOR. The Pt/GA-CNx-ZIF-8 catalyst was found to have an ECSA value of 89.0 m^2 g^{-1}_{Pt}, much higher than its GA predecessors. This signified higher utilization of Pt with a higher peak current density of 0.68 A mg^{-1}_{Pt} and high temperament towards MOR showing only a slight reduction even after 1000 cycles. This points out that the support offers higher resistance towards corrosion preventing agglomeration and migration of Pt due to anchoring effect of N species on Pt.

The catalyst also indicated resistance towards CO poisoning by showing an ultrahigh constant stable potential. Fig. 6 [42] shows the process of synthesis of 3D boron nitrogen co-doped GA.

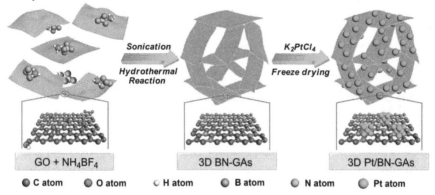

Fig. 6 Synthetic processes for the 3D Pt/BN-GA architecture. [Reprinted with permission from Ref. [42]. Copyright (2018) American Chemical Society].

Xiaoteng Liu et al. [43] demonstrated a new GA based DMFC design by replacing the gas diffusion layer (GDL) and the flow field plate. The PtRu based GA electrocatalyst along with the novel assembly showed a maximum power density of 24.95 mW cm^{-2} and a mass power density of 5.02 Wkg^{-1} which was way better than any ordinary DMFC. This assembly showed a possibility to use highly concentrated fuels as it witnessed suppression in methanol crossover with 12 M methanol.

A 3D porous holey GA was synthesized by X. Zhang et al. [44] and the electrocatalyst, 0.5h Pt/HGA showed ESA of 88.35 m^2 g^{-1}. This larger ESCA was due to better dispersion of fine PtNPs.

ZhouXu et al. [45] developed a flower-like nanostructured V$_3$S$_4$/graphene aerogel (V$_3$S$_4$-GA) hybrid which contained a large surface area with abundance of active sites for catalyst binding providing an excellent network for electron transportation. The V$_3$S$_4$/GA hybrid showed a high electron transfer number and high current density of about 3.97 and 9.5 mA cm^{-2} respectively along with high resistance towards MOR displaying good durability. After 30,000s at a constant voltage, the V$_3$S$_4$/GA showed a decrement of only 18%.

N-GA was synthesized by QiXue et al. [46] via pyrolysis of graphene aerogels-polyallylamine (GA-PAA) with the variable nitrogen content, depending upon the molecular weight of PAA. The higher electrocatalytic properties of this catalyst lie in the 3D interconnected porous structure with activity and durability depending directly on the variation of the Nitrogen content. Unlike the commercial Pt/C, N-GA exhibited resistance to CO poisoning. After 2000 cycles, the onset potential suffers only a small decrement indicating superior stability.

2.3   Doped aerogels

To surpass the issue related to poor durability, tin dioxide was replaced with carbon black as an catalyst support in PEMFCs by G. Ozouf et al. [47] where in, they developed doped SnO_2 aerogels that not only had high electronic conductivity but also an adaptable structure. Sol-gel technique was employed to synthesis SnO_2 xerogels and aerogels using metal alkoxides as precursors. The material exhibited surface area of 80-90m^2/g with bimodal pore size distribution centered on around 25 and 45 nm. The electronic conductivity of doped SnO_2 (1 S/cm) was comparable with that of Vulcan XC-72 (4 S/cm). Thus, making way for oxides as an alternative to be used as an electrocatalyst in energy storage devices. Weijian Yuan et al. [48] synthesized nitrogen-doped CA using resorcinol, formaldehyde and GO and carbonization of the aforementioned precursors and hydrogel under ammonia obtaining an ECSA value of 41.57 m^2 g^{-1} and a peak power density of 22.41 mW cm^{-2}. The aerogel showed a very high surface area and superior doping of N and O atoms. This lead to high hydrophilicity and strong water absorption. The two layered-cathode designed improved the electrocatalytic properties by decreasing cathode polarization followed by an increment in anode polarization. Construction of a water management layer (WML) for DMFC using the as developed aerogel relieved it of water flooding by expressing a stable output voltage during constant current density discharges.

Miaomiao Li et al. [42] presented a bottom up approach to fabricate 3D boron and nitrogen co- doped graphene aerogels (BN-GAs) using a hydrothermal process to form doped graphene hydrogels which were later freeze-dried to form BN-GA. The Pt based electrocatalyst contained cross-linked pores leading to high specific surface areas and an abundance of boron and nitrogen active sites. This resulted in ECSA value of up to 106.0 m^2 g^{-1} which is much larger than its commercial predecessors. The electrode potential was also found out to be lower implying that the Pt/BN-GA electrocatalyst facilitates methanol oxidation. Durability tests show that after 2000s there was only a decrement of 39% in the initial current with negligible changes the morphology. Fig. 7 [42] shows

SEM and TEM (Transmission electron microscopy) images of Pt NPs dispersed over BN-GAs.

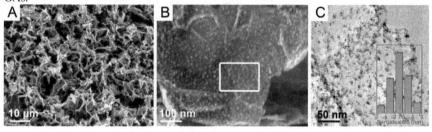

Fig. 7 Typical (A and B) FE-SEM and (C) TEM images show that small Pt NPs are dispersed uniformly on 3D porous BN-GA supports [Reprinted with permission from Ref. [42]. Copyright (2018) American Chemical Society].

Nitrogen doped 3D GA was fabricated by LeiZhao et al. [49] using supramolecular assembly-assisted method which lead to in situ prevention of re-stacking along with pore formation and subsequent nitrogen doping. As the supramolecular aggregates direct the structure formation of the aerogel, the specific surface area along with stacking properties and porosity depend on it, which ultimately leads to durability, stability and tolerance to poisoning. Having an ECSA value of 60.6 m^2 g^{-1} and a current density of 507.5 mA mg^{-1}, which is much better than its commercial counterpart and it was concluded that Pt/NGA possessed superior catalytic activity. After 1000 cycles, the Pt/NGA electrode only suffered 25% loss in its initial current density with very little growth in the sizes of the Pt NPs showing that the catalyst possessed good stability.

Another study by M. Selim Çögenli et al. [50] discusses nitrogen and boron hetero-doped Gas for Pt based electrocatalyst which would further improve electrocatalytic activities for MOR. Their study with various catalysts revealed the information given in table 1 [50].

The below table reveals high stability of Pt/BGA towards MOR due to strong interaction between the catalyst and the doped graphene support along with 3D interconnected frameworks having a macroporous structure. However, N doped graphene showed higher tolerance towards CO poisoning for MOR, which indicates doped GA as a promising candidate.

Table 1 Electrocatalytic properties of various electrocatalysts [50].

Electrocatalysts	BET specific surface area (m^2 g^{-1})	final and initial current density ratios	I_f(mA/cm^2)	I_b(mA/cm^2)	I_f/I_b
Pt/C	-	0.19	11.98	11.64	1.03
Pt/GA	244.91	0.06	8.72	4.32	2.02
Pt/NGA	379.71	0.20	9.42	4.63	2.03
Pt/BGA	-	0.41	19.28	18.31	1.05

2.4   Mesoporous carbon

A highly conducting mesoporous carbon (MC) aerogel was synthesized by RashmiSingh et al. [51] wherein high temperature and high-pressure gelation of resorcinol with furfuraldehyde and consequent ambient drying and carbonization resulted in a mesoporous structure with high electrical conductivity. Pt/CA catalyst exhibited superior ECSA, more positive onset potential (964 mV) and half wave potential (814 mV) towards ORR kinetics owing to the Pt particles of size ~ 3nm that are uniformly dispersed on the highly conducting MC aerogels support. A power density of 536 mW cm^{-2} (0.6 V) at 60°C was found for the Pt/CA showing its potential as electro-catalyst for PEMFC applications. Fig. 8 [52] shows ~~an image of Pt/CA catalyst obtained via TEM~~a TEM image of Pt/CA prepared by an ethylene glycol (EG; 99.5%) liquid-phase reduction method.

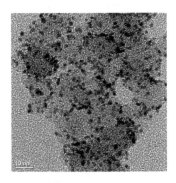

Fig.8 TEM Image of Pt/CA catalyst[Reprinted with permission from Ref. [52].
Copyright (2009) American Chemical Society].

A study showed that mesoporous CA when impregnated with ceramic metal oxides like titania and ceria have greater stability. P. Kolla et al. [53] synthesized PtRu based mesoporous carbon electrocatalyst via a modified sol-gel Pechini method. The two complexes TiO_2-PtRu and CeO_2-PtRu showed electronic interaction between ruthenium and pure metal oxide participated in the removal of carbonaceous species. While TiO_2-PtRu/CA showed greatest stability in acid solutions, the CeO_2-PtRu/CA did the same in basic solutions. Carmen I. Fort et al. [54] also developed Pt-Ru based highly mesoporous CAs using sol-gel method which showed a specific area of 843 m^2/g and a maximum current density of 153.1 mA/cm^2.

3. Non-precious catalyst using aerogels as support for DMFC applications

Since, the reactions involved in a FC are highly corrosive in nature, precious metals are used as catalyst for applications which are very costly. Hence, an ongoing research over non-precious catalysts and supports show considerable promise for further future developments.

Palladium aerogels modified with α-, β-, or γ-cyclodextrins (Pd_{CD})were synthesized by self-assembly method by Wei Liu et al. [55] which gave an ECSA of 98 m^2 g^{-1} and an I_f/I_b ratio of 1.16. Along with superior properties, it also showed high polarization current densities with much slower decay. SanthanaSivabalanJayaseelan et al. [56] developed a non-precious electrocatalyst as a cost-effective approach of replacing Pt based commercial electrocatalysts. A sol-gel technique was employed for the synthesis of $NiCo_2O_4$-MWCNT nanocomposite aerogels. The addition of MWCNTs increases the current densities with a peak current (41.42 mA cm^{-2}) of ethanol oxidation. After 1000 continuous potential cycles only 17% loss was noted indicating good cyclic stability.

Conclusionand outlook

The main focus of this book chapter was on reporting different types of aerogels that are utilized as catalyst support, mostly carbon based for improving the ORR and MOR activity of the FC which is primarily one of the reasons hindering the commercialization of FC on larger scale. There are tons of literatures available on several types of aerogels for various application ranging from energy storage devices to catalyst support but the research in this field even after a century is still in infancy. Despite being an active area of research, there are a lot of hurdles that are blocking their use on a larger scale. The supports developed so far could not exhibit high-surface area, porosity, conductivity, resistance to corrosion and prevent water flooding all at ones with the same catalyst support [57]. Catalyst with improved ORR kinetics, electrochemical catalytic activity manufactured at a lower cost with a higher performance is an urgent need for various sustainable energy storage applications.

Despite its great performance, the use of aerogels for FC electrocatalysts is met with many complications. The first and foremost being the difficulty in controlling the physiochemical and mechanical properties precisely during synthesis [58]. There is a limited class of aerogels available and thus synthesis of novel aerogels is in high demand. Even though new advancements in the field are happening, aerogels are met with a major difficulty and that is, its production cannot be scaled up industrially. The technology required for mass production is very expensive and it cannot be generalized for all kinds of aerogels. FCs demanding novel and more catalytically active aerogels require much more expensive methods for doping, co-doping and altering 3-D structures, etc.

References

[1] R. Jasinski, A new FC cathode catalyst, Nature 201 (1964) 1212−1213. https://doi.org/10.1038/2011212a0

[2] N. Cheng, M.N.Banis, J. Liu, A. Riese, X. Li, R. Li, S. Ye, S. Knights, X. Sun, Extremely stable platinum nanoparticles encapsulated in a zirconia nanocage by area-selective atomic layer deposition for the oxygen reduction reaction, Adv. Mater. 27 (2015) 277−281. https://doi.org/10.1002/adma.201404314

[3] B.Y. Xia, H.B. Wu, N. Li, Y.Yan, X.W. Lou, X.Wang, One pot synthesis of Pt−Co alloy nanowire assemblies with tunable composition and enhanced electrocatalytic properties, Angew. Chem. Int. Ed. 54 (2015) 3797−3801. https://doi.org/10.1002/anie.201411544

[4] D. Higgins, M.A. Hoque, M.H.Seo, R. Wang, F. Hassan, J.Y. Choi, M. Pritzker, A. Yu, J. Zhang, Z. Chen, Development and simulation of sulfur-doped graphene supported platinum with exemplary stability and activity towards oxygen reduction, Adv. Funct. Mater. 24 (2014)4325−4336. https://doi.org/10.1002/adfm.201400161

[5] X. Fu, J.Y. Choi, P. Zamani, G. Jiang, M.A. Hoque, F.M. Hassan, Z. Chen, Co−N Decorated hierarchically porous graphene aerogel for efficient oxygen reduction reaction in acid, ACS Appl. Mater. & Interfaces 8 (2016) 6488−6495. https://doi.org/10.1021/acsami.5b12746

[6] S.S. Kistler, Coherent expanded aerogels and jellies, Nature 127 (1931) 741-741. https://doi.org/10.1038/127741a0

Materials Research Forum LLC
https://doi.org/10.21741/9781644901298-5

[7]P.C.Baena, A.G. Agrios, Transparent conducting aerogels of antimony-doped tin oxide, ACS Appl. Mater. Interfaces 6 (2014) 19127-19134. https://doi.org/10.1021/am505115x

[8]I.U. Arachchige, S.L. Brock, Sol-Gel assembly of CdSe nanoparticles to form porous aerogel networks, J. Am. Chem. Soc. 128 (2006) 7964-7971. https://doi.org/10.1021/ja061561e

[9]M.A. Worsley, P.J. Pauzauskie, T.Y. Olson, J. Biener, J.H. Satcher, Jr.,T.F. Baumann, Synthesis of graphene aerogel with high electrical conductivity, J. Am. Chem. Soc. 132 (2010) 14067-14069. https://doi.org/10.1021/ja1072299

[10]H.Sun, Z. Xu, C. Gao, Multifunctional, ultra-flyweight, synergistically assembled carbon aerogels, Adv. Mater. 25 (2013) 2554-2560. https://doi.org/10.1002/adma.201204576

[11]M.A. Worsley, S.O. Kucheyev, H.E. Mason, M.D. Merrill, B.P. Mayer,J. Lewicki, C.AValdez, M.E. Suss, M. Stadermann, P.J. Pauzauskie, J.H. Satcher Jr., J. Bienera, T.F. Baumanna, Mechanically robust 3D graphene macroassembly with high surface area, Chem. Commun. 48 (2012) 8428-8430. https://doi.org/10.1039/C2CC33979J

[12]T.F.Baumann, M.A.Worsley, T.Y.J.Han, J.H.SatcherJr., High surface area carbon aerogel monoliths with hierarchical porosity, J. Non-Cryst. Solids 354 (2008) 3513-3515. https://doi.org/10.1016/j.jnoncrysol.2008.03.006

[13]W.Liu, A.K. Herrmann,N.C. Bigall, P. Rodriguez, D. Wen, M. Oezaslan, T.J. Schmidt, N.Gaponik,A. Eychmüller, Noble metal aerogels-synthesis, characterization, and application as electrocatalysts, Acc. Chem. Res. 48 (2015)154–162. https://doi.org/10.1021/ar500237c

[14]A. Rabis, P. Rodriguez,T.J. Schmidt, Electrocatalysis for polymer electrolyte FCs: recent achievements and future challenges, ACS Catal. 2 (2012) 864−890. https://doi.org/10.1021/cs3000864

[15]C.H. Cui, S.H. Yu, Engineering interface and surface of noble metal nanoparticle nanotubes toward enhanced catalytic activity for fc applications, Acc. Chem. Res. 46 (2013)1427−1437. https://doi.org/10.1021/ar300254b

[16]Z. Chen, M.Waje, W. Li, Y. Yan, Supportless Pt and PtPd nanotubes as electrocatalysts for oxygen-reduction reactions, Angew. Chem. Int. Ed. 46 (2007) 4060−4063. https://doi.org/10.1002/anie.200700894

[17]K. Shehzad, Y. Xu, C. Gao, X. Duan, Three-dimensional macro-structures of two-dimensional nanomaterials, Chem. Soc. Rev. 45 (2016) 5541-5588. https://doi.org/10.1039/C6CS00218H

[18]C. Hu, D. Liu, Y. Xiao, L. Dai, Functionalization of graphene materials by heteroatom-doping for energy conversion and storage, Prog. Nat. Sci.28 (2018) 121–132. https://doi.org/10.1016/j.pnsc.2018.02.001

[19]J. Mao, J.Iocozzia, J. Huang, K. Meng, Y. Lai, Z. Lin, Graphene aerogels for efficient energy storage and conversion, Energ. & Environ. Sci. 11 (2018) 772–799. https://doi.org/10.1039/C7EE03031B

[20]USDOE (Department of Energy) https://www.energy.gov/sites/prod/files/2017/11/f46/FCTT_Roadmap_Nov_2017_FIN AF.pdf (last accessed on 13 Jan 2020)

[21]H. Zhu, Z. Sun, M. Chen, H. Cao,K. Li, Y. Cai, F. Wang, Highly porous composite based on tungsten carbide and N-doped carbon aerogels for electrocatalyzing oxygen reduction reaction in acidic and alkaline media, Electrochim. Acta 236 (2017) 154–160. https://doi.org/10.1016/j.electacta.2017.02.156

[22]W. Yuan, C. Hou, X. Zhang, S. Zhong, Z. Luo, D. Mo, Y. Zhang, X. Liu, Constructing cathode catalyst layer of a passive direct methanol FC with highly hydrophilic carbon aerogel for improved water management, ACS Appl. Mater. Interfaces 11 (2019)37626-37634. https://doi.org/10.1021/acsami.9b09713

[23]H.Zhu, Z. Guo, X. Zhang, K.Han, Y. Guo, F. Wang, Z. Wang, Y. Wei, Methanol-tolerant carbon aerogel-supported Pt–Au catalysts for direct methanol FC, Int. J. Hydrogen Energy37 (2012) 873–876. https://doi.org/10.1016/j.ijhydene.2011.04.032

[24]S. Bong, D. Han, Mesopore-controllable carbon aerogel and their highly loaded PtRu anode electrocatalyst for DMFC applications, Electroanalysis 32 (2019) 104-111. https://doi.org/10.1002/elan.201900320

[25]W. Xia, C. Qu, Z. Liang, B. Zhao, S. Dai, B.Qiu, Y. Jiao, Q. Zhang, X. Huang, W.Guo, D. Dang, R. Zou, D. Xia, Q. Xu, M. Liu, High-performance energy storage and conversion materials derived from a single metal–organic framework/graphene aerogel composite, Nano Lett. 17 (2017) 2788–2795. https://doi.org/10.1021/acs.nanolett.6b05004

[26]Y. Wang, L. Zou, Q. Huang, Z.Zou, H.Yang, 3D carbon aerogel-supported PtNi intermetallic nanoparticles with high metal loading as a durable oxygen reduction

electrocatalyst, Int.J. Hydrogen Energy 42 (2017) 26695–26703.
https://doi.org/10.1016/j.ijhydene.2017.09.008

[27]W. Yuan, X. Zhang, Y. Zhang, X. Liu, Improved anode two-phase mass transfer
management of direct methanol FC by the application of graphene aerogel, ACS
Sustainable Chem. Eng. 7 (2019)11653-11661.
https://doi.org/10.1021/acssuschemeng.9b01665

[28]S. Zhao, H. Yin, L. Du, G. Yin, Z. Tang, S. Liu, Three dimensional N-doped
graphene/PtRu nanoparticle hybrids as high performance anode for direct methanol
FCs, J. Mater. Chem. A 2 (2014) 3719-3724. https://doi.org/10.1039/C3TA14809B

[29]J.Duan, X. Zhang, W. Yuan, H. Chen, S. Jiang, X.Liu, Y. Zhang, L. Chang, Z. Sun,J.
Du, Graphene oxide aerogel-supported Pt electrocatalysts for methanol oxidation, J.
Power Sources 285 (2015) 76–79. https://doi.org/10.1016/j.jpowsour.2015.03.064

[30]M. Liu, C. Peng, W. Yang, J. Guo, Y. Zheng, P. Chen, T. Huang, J. Xu, Pd
nanoparticles supported on three-dimensional graphene aerogels as highly efficient
catalysts for methanol electrooxidation, Electrochim. Acta 178 (2015) 838–846.
https://doi.org/10.1016/j.electacta.2015.08.063

[31]L. Zhao, Z.B. Wang, J.L. Li, J.J. Zhang, X.L. Sui, L.M. Zhang, Hybrid of carbon-
supported Pt nanoparticles and 3Dgraphene aerogel as high stable electrocatalyst for
methanol electrooxidation, Electrochim. Acta 189 (2016) 175–183.
https://doi.org/10.1016/j.electacta.2015.12.072

[32]X. Zhang, N. Hao, X. Dong, S. Chen, Z. Zhou, Y. Zhang, K. Wang, One-pot
hydrothermal synthesis of platinum nanoparticle-decorated three-dimensional
nitrogen-doped graphene aerogel as a highly efficient electrocatalyst for methanol
oxidation, RSC Adv. 6 (2016) 69973–69976. https://doi.org/10.1039/C6RA12562J

[33]X. Peng, D. Chen, X. Yang, D. Wang, M. Li, C.C. Tseng, R.Panneerselvam, X.
Wang, W. Hu, J. Tian, Y. Zhao, Microwave-assisted synthesis of highly dispersed
PtCu nanoparticles on three-dimensional nitrogen-doped graphene networks with
remarkably enhanced methanol electrooxidation,ACS Appl. Mater.
Interfaces8(2016)33673–33680. https://doi.org/10.1021/acsami.6b11800

[34]L. Zhao, X.L. Sui, J.L. Li, J.J. Zhang, L.M.Zhang, Z.B. Wang, Ultra-fine Pt
nanoparticles supported on 3D porous N-doped graphene aerogel as a promising
electro-catalyst for methanol electrooxidation, Catal. Commun. 86 (2016) 46–50.
https://doi.org/10.1016/j.catcom.2016.08.011

Materials Research Forum LLC
https://doi.org/10.21741/9781644901298-5

[35] Y.H. Kwok, Y.F. Wang, A.C.H. Tsang, D.Y.C. Leung, Ru@Pt core shell nanoparticle on graphene carbon nanotube composite aerogel as a flow through anode for direct methanol microfluidic FC, Energy Procedia 142 (2017) 1522–1527. https://doi.org/10.1016/j.egypro.2017.12.602

[36] H. Huang, J. Zhu, D. Li, C. Shen, M. Li, X.Zhang, Q. Jiang, J. Zhang, Y. Wu, Pt nanoparticles grown on 3D RuO$_2$-modified graphene architectures for highly efficient methanol oxidation, J. Mater. Chem. A 5 (2017)4560–4567. https://doi.org/10.1039/C6TA10548C

[37] Y.H. Kwok, A.C.H. Tsang, Y.Wang, D.Y.C. Leung, Ultra-fine Pt nanoparticles on graphene aerogel as a porous electrode with high stability for microfluidic methanol FC, J. Power Sources 349 (2017) 75–83. https://doi.org/10.1016/j.jpowsour.2017.03.030

[38] L.M. Zhang, X.L. Sui, L. Zhao, G.S. Huang, D.M. Gu, Z.B. Wang, Three-dimensional hybrid aerogels built from graphene and polypyrrole-derived nitrogen-doped carbon nanotubes as a high-efficiency Pt-based catalyst support, Carbon 121 (2017) 518–526. https://doi.org/10.1016/j.carbon.2017.06.023

[39] Y. Zhou, X.C.Hu, S. Guo, C. Yu, S.Zhong, X. Liu, Multi-functional graphene/carbon nanotube aerogels for its applications in supercapacitor and direct methanol FC, Electrochim. Acta 264 (2018) 12–19. https://doi.org/10.1016/j.electacta.2018.01.009

[40] M. Yan, Q. Jiang, T. Zhang, J. Wang, L. Yang, Z. Lu, H. He, Y. Fu, X. Wang, H. Huang, Three-dimensional low-defect carbon nanotube/nitrogen-doped graphene hybrid aerogel-supported Pt nanoparticles as efficient electrocatalysts toward methanol oxidation reaction, J. Mater. Chem. A 6(2018)18165-18172. https://doi.org/10.1039/C8TA05124K

[41] X.-L. Sui, L.M. Zhang, L. Zhao, D.M. Gu, G.S. Huang, Z.-B. Wang, Nitrogen-doped graphene aerogel with an open structure assisted by in-situ hydrothermal restructuring of ZIF-8 as excellent Pt catalyst support for methanol electro-oxidation, Int. J. Hydrogen Energy 43(2018)21899-21907. https://doi.org/10.1016/j.ijhydene.2018.09.223

[42] M. Li, Q. Jiang, M. Yan, Y. Wei, J.Zong, J. Zhang, Y. Wu,H. Huang, Three-Dimensional boron- and nitrogen-codoped graphene aerogel-supported Pt nanoparticles as highly active electrocatalysts for methanol oxidation reaction, ACS sustain. 6 (2018) 6644–6653. https://doi.org/10.1021/acssuschemeng.8b00425

[43] X. Liu, J. Xi, B.B. Xu, B. Fang, Y. Wang, M.Bayati, K. Scott, C. Gao, A High-performance direct methanol FC technology enabled by mediating high-concentration

Materials Research Forum LLC
https://doi.org/10.21741/9781644901298-5

methanol through a graphene aerogel, Small methods 2(2018) 1800138. https://doi.org/10.1002/smtd.201800138

[44]X. Zhang, L.Zhou, Y.Whang, J. Tang, J. Li, Facile synthesis of holey graphene-supported Pt catalysts for direct methanol electro-oxidation, Microporous Mesoporous Mater.247(2017)116-123. https://doi.org/10.1016/j.micromeso.2017.03.061

[45]Z. Xu, Y. Zhang, Y. Wang, L. Zhan, Flower-like nanostructured V_3S_4 grown on three-dimensional porous graphene aerogel for efficient oxygen reduction reaction, App. Surf. Sci. 450 (2018) 348–355. https://doi.org/10.1016/j.apsusc.2018.04.163

[46]Q.Xue, Y. Ding, Y.Xue, F. Li,P. Chen, Y.Chen, 3D nitrogen-doped graphene aerogels as efficient electrocatalyst for the oxygen reduction reaction, Carbon 139 (2018) 137–144. https://doi.org/10.1016/j.carbon.2018.06.052

[47]G. Ozouf, C. Beauger, Niobium- and antimony-doped tin dioxide aerogels as new catalyst supports for PEM FCs, J. Mater. Sci. 51 (2016) 5305–5320. https://doi.org/10.1007/s10853-016-9833-7

[48]W. Yuan, C. Hou, X. Zhang, S. Zhong, Z. Luo, D. Mo, Y. Zhang,X. Liu, Constructing cathode catalyst layer of a passive direct methanol FC with highly hydrophilic carbon aerogel for improved water management, ACS Appl. Mater. Interfaces 11 (2019) 37626-37634. https://doi.org/10.1021/acsami.9b09713

[49]L. Zhao,X.-L. Sui, J.Z. Li, J.J. Zhang, L.M. Zhang,G.S. Huang, Z.B. Wang,Supramolecular assembly promoted synthesis of three-dimensional nitrogen doped graphene frameworks as efficient electrocatalyst for oxygen reduction reaction and methanol electrooxidation, Appl. Catal. 231 (2018) 224–233. https://doi.org/10.1016/j.apcatb.2018.03.020

[50]M.S. Çögenli, A.B. Yurtcan,Heteroatom doped 3D graphene aerogel supported catalysts for formic acid and methanol oxidation, Int J. Hydrogen Energy 45 (2019) 650-666. https://doi.org/10.1016/j.ijhydene.2019.10.226

[51]R. Singh, M.K. Singh, S. Bhartiya, A. Singh, D.K. Kohli, P.C. Ghosh, S. Meenakshi, P.K. Gupta, Facile synthesis of highly conducting and mesoporous carbon aerogel as platinum support for PEM FCs, Int. J. Hydrogen Energy 42 (2016) 11110–11117. https://doi.org/10.1016/j.ijhydene.2017.02.207

[52]S. Wei, D. Wu, X. Shang,R. Fu, Studies on the structure and electrochemical performance of Pt/Carbon aerogel catalyst for direct methanol FCs, Energy Fuels 23 (2009) 908–911. https://doi.org/10.1021/ef8006432

[53]P. Kolla, K. Kerce, Y. Normah, H. Fong, A. Smirnova, Metal oxides modified mesoporous carbon supports as anode catalysts in DMFC, ECS Trans. 45 (2013) 35–45. https://doi.org/10.1149/04521.0035ecst

[54]C.I. Fort, L.C. Cotet, F.Vasiliu, P.Marginean, V.Danciu, I.C. Popescu,Methanol oxidation at carbon paste electrodes modified with (Pt–Ru)/carbon aerogels nanocomposites,Mater. Chem. Phys.172 (2016) 179–188. https://doi.org/10.1016/j.matchemphys.2016.01.061

[55]W. Liu, A.K. Herrmann, D. Geiger,L. Borchardt,F. Simon,S.Kaskel, N.Gaponik, A.Eychmüller, Ahigh-performance electrocatalysis on palladium aerogels, Angew. Chem. Int. Ed. 51 (2012) 5743–5747. https://doi.org/10.1002/anie.201108575

[56]S.S.Jayaseelan, S. Radhakrishnan,B.Saravanakumar, M.K.Seo, M.S. Khil, H.Y.Kim, B.S. Kim,Novel MWCNT interconnected $NiCo_2O_4$ aerogels prepared by a supercritical CO_2 drying method for ethanol electrooxidation in alkaline media, Int. J. Hydrogen Energy 41 (2016) 13504–13512. https://doi.org/10.1016/j.ijhydene.2016.05.175

[57]S.Shahgaldi, J. Hamelin, Improved carbon nanostructures as a novel catalyst support in the cathode side of PEMFC: a critical review, Carbon 94 (2015) 705–728. https://doi.org/10.1016/j.carbon.2015.07.055

[58]L.Zuo, Y. Zhang, L. Zhang, Y.E. Miao,W.Fan,T. Liu, Polymer/carbon-based hybrid aerogels: preparation, properties and applications, Materials (Basel, Switzerland) 8 (2015) 6806–6848. https://doi.org/10.3390/ma8105343

Aerogels II: Preparation, Properties and Applications   Materials Research Forum LLC
Materials Research Foundations 97 (2021) 99-120   https://doi.org/10.21741/9781644901298-6

Chapter 6

Aerogels Utilizations in Batteries

T. Pazhanivel[1]*, S. Dhinesh[2], M. Priyadharshini[1], R.Gobi[2]

[1] Department of Physics, Periyar University, Salem 636 011, Tamilnadu, India

[2] PG and Research Department of Physics, Arignar Anna Government Arts College, Namakkal-02, Tamilnadu, India

* pazhanit@gmail.com

Abstract

Aerogels, a nanoscale 3D mesoporous spongy sample of enhanced surface area, was usually considered as insulator for thermal application, catalyst, and as radiation detector. Presently, it is investigated as potential candidate for electrochemistry due to its inborn capacity to enhance the characteristic features of the surfaces of commercial active materials in batteries and ultracapacitors. Recently composite aerogels which is blended with metal oxides, metal sulphides and so on have been set up as low thickness, profoundly permeable, and large amount of accessible surface and examined as active electrodes. This type of aerogel-based composites challenges the standard manners by that electrochemically active materials are considered, examined, and employed.

Keywords

Aerogel, Electrochemical Batteries, Specific Capacity, Coulombic Efficiency, Rate Capability

Contents

1.   Introduction

In this modern era, the plenteous data around people is getting significant comfort. On this aspect enormous requests for the opportune what's more, productive conveyance of worldwide data, and its related process for efficient application demands a portable device for real world application [1]. The electronic gadgets comprising cellular, laptops, and flexible electronic systems are the competitors which has advanced a quick improvement in data preparation and distribution on the improvement and development of modern technologies. Thus portable electronics devices (PED) have been quickly becoming potential candidates in the recent past [2-5]. The essential inspiration driving such movement is, PED were generally utilized from individual systems to highly-innovation systems for example in aerospace because of its capacity in incorporation and association with the modern life system, that may buy incredible comfort and significant changes, even may turn into a irreplaceable component of all individual operations. On behalf of developing demands of high power PEDs, the ability of energy frameworks have to be redesigned. As needs be, investigating efficient, long-life, safe, and huge limit energy storing devices are the highly difficult challenges of PEDs [6,7].

Electrochemical energy storage system, particularly energize capable batteries, is broadly utilized as the source of energy in PEDs towards a considerable amount of time and advanced the successful development of PED. To fulfill the consistently high necessities of PEDs, important enhancements battery technology has to be accomplished. Battery technology in PEDs have experienced a vast variety of batteries like lead- corrosive, nickel-cadmium (Ni-Cd), nickel-metal hydride (Ni-MH), lithium-particle (Li-particle) and so on. Its energy density and explicit force were significantly mended over period. However, the present battery innovation can't completely make up for lost time with the fast development of electronic devices [3,8-10]. The electronic devices has exposed a numerous disadvantages, that is, restricted energy storing limit such as its low stability, and considerably more self-release, that turned a lagging point for the advancement in

Aerogels II: Preparation, Properties and Applications   Materials Research Forum LLC
Materials Research Foundations **97** (2021) 99-120   https://doi.org/10.21741/9781644901298-6

electronic devices. Generally, the powerful utilization of multifunctional electronic devices anticipates energy related frameworks with better energy, weightless, and highly stable. In any case, the present energy systems try to fulfill the regularly expanding requests of rising hardware based fields. In this manner, the normal plan and generation of novel batteries has a persevering sought after objective for the future electronic devices. Gigantic endeavors have been devoted to enhancing the redox attributes of electrochemical system [11]. Notable improvement was achieved by late literatures. They constitutes additionally various surprising surveys that spread the master gress of battery technologies. Thinking about the basic commitment of battery advancements to the improvement of electronic devices, it is of extraordinary passion to condense the advancement of battery for electronic devices in the previous decades.

The significance of energy systems for advanced portable electronic devices is discussed. Four kinds of rechargeable batteries with an explanation of advancement in various kinds of PED were portrayed elaborately. Specific concern inclines to those regular PED, for example, cellular, tablets, computerized camera, just like the recently developing PEDs and counting wearable electronic gadgets. At long last, the present advancement patterns of the battery technologies and the future electronic devices have been discussed.

Different electrochemically dynamic materials have been set up through sol-gel forms going from gel electrolyte to a few distinctive type intercalation electrodes. Applying a suitable alteration in the system, especially evacuation of the porosity, active samples were set up as powders and thin films coated on substrates. As of late, aerogel based samples were presented as promising positive electrode for Li-ion based system [12,13]. Such materials, acquired by a supercritical drying, are described by a continuous architecture of solid phase materials side by side its pores. The dimension of the solid phase around the pores ranges between 10 to 30 nm. This sort of morphology is related with an exceptionally high surface region. The redox behaviour of aerogel is clearly changed by such a peculiar structure. The porosity enables the facile penetration of electrolyte deep into the aerogel. Likewise, the sample possess an exceptionally thin-film of solid phase which again reduces the length of dispersion ways for the intercalated particles. Such a mixing aerogel cathodes to a great extent unaffected by dispersion restrictions on lithium addition.

2. Types of batteries

Batteries are the energy storage device that store the energy through the electrochemical reactions. The batteries are irreplaceable devices that have highest energy density around $120 - 200$ whkg^{-1}, when compared to the supercapacitors. There are two types of batteries namely primary and the secondary batteries.

The primary batteries are used once because of irreversible reaction like alkaline battery. On the other hand the secondary batteries are rechargeable because of reversible reactions. In compared to primary batteries, secondary batteries are consider as a potential candidate for portable electronic devices. The secondary batteries include lead - acid battery, metal-ion battery and metal – air battery. The different types of batteries were summarized in fig. 1 [12].

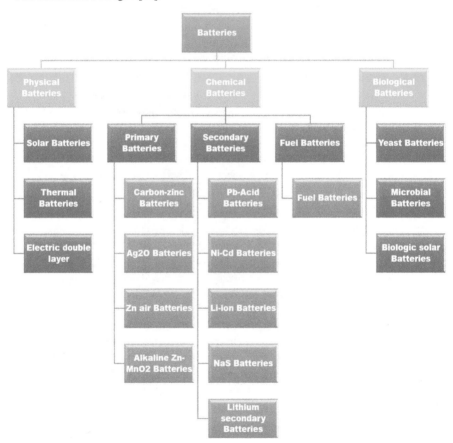

Fig. 1 *Types of batteries [12].*

Aerogels II: Preparation, Properties and Applications   Materials Research Forum LLC
Materials Research Foundations **97** (2021) 99-120   https://doi.org/10.21741/9781644901298-6

2.1   Lead - acid battery

Lead acid battery was the first rechargeable battery for the energy storage application. The efficiency of the device range from few watt hours to megawatt hour which successfully extents its wide application from a common household up to load leveling submarine power. The major advantage of the lead acid battery is reliability, wide range of operating temperature, high current density, high abundance of raw materials and relatively low cost for the manufacturing process. The advanced lead acid batteries have been used in smaller scale domestic and commercial energy storage application.

The sulphuric acid dissolves, the molecules break up into negative sulphate ions (So_4^-) and positive hydrogen ions ($2H^+$). The two electrodes are connected to the DC supply and immersed in the solution. One electrode moves towards the positive hydrogen ions and connected to the negative terminal of the supply and another electrode move towards the negative sulphate ions and connected to the positive terminal of the supply. The hydrogen ions and sulphate ions react with water and form sulphuric acid. The voltmeter is connected between the two electrodes and it shows potential difference between them at the DC power supply is disconnected. When the DC power supply is connected to the electrode, the current will flow from positive electrode to the negative electrode through external circuit and the cell is capable of supplying electrical energy [14].

2.2   Metal - ion battery

The metal – ion battery includes Li – ion battery, semi- solid Li – ion battery, aluminum – ion battery and redox polymer battery. The metal ion battery have the excellent energy storage, no memory effect, long cyclability and low self discharge. Among the various metal - ion battery, the Lithium – ion battery has the better performance. The Li – ion battery have an high energy density and power density. It is used for the portable electronic devices, power tools and hybrid vehicles. The redox reaction of the $LiMn_2O_4$ is

Positive electrode

$$Li_2MnO_3.LiMo_2 \rightarrow LiMnO_3.Mo_2 + Li^+ + e^-$$

Negative electrode

$$6C + xLi^+ + xe^- \Leftrightarrow C_6Li_x$$

The anode and cathode of the LIBs as an active component of the charge / discharging process involves the intercalation and dis-intercalation of the Li ions. When the battery is charging from the metal ion electrode, the lithium ions of cathode pass through the electrolyte to the negative and graphite electrode. During the process the battery stores the energy. In converse during discharging, the Li ions move come back to the positive

Materials Research Forum LLC
https://doi.org/10.21741/9781644901298-6

electrode through the electrolyte and e⁻ through outer connection. Thus produces the required energy [15].

2.3   Metal air battery

The metal air battery (MABs) operates in an open air battery. It consists of metal anodes and air cathodes. The MABs are cheap because of cathode (oxygen from air) and the anodes can be produced at low cost metals such as Zn, Fe, Mg, Li, Na, K and Al. The performance of various battery configuration is to depend on the theoretical energy density. The lithium air battery have the highest theoretical energy density (5928 whkg⁻¹) and high cell potential (2.96 V) when compared to the other metal air batteries. The iron air battery has the smallest theoretical energy density (1080 whkg⁻¹) and low cell potential (1.28 V).

The major advantage of MABs in many application is, the cathode uses oxygen from ambient air, which leads to weight reduction of the battery. The MABs have short life cycle, less safety and low rate capability are the major disadvantages of the MABs. The application of MABs as the energy storage have been used as a various technologies. The theoretical performance limit of few metals were listed in the table 1. The mechanism of metal air battery can be different from the traditions ionic batteries. In MABs, the metals or alloys transform the metallic ions at the anode to the cathode of oxygen transform to hydroxide ions. The metal releases electrons, transform to metallic ions and dissolves into electrolyte during the aqueous and non – aqueous electrolyte. These process will be the charging procedure of the MABs [16].

Table 1   Theoretical specific energies, energy densities, and operating battery voltages of various metal–air batteries (MABs).

Metals	Specific energy (wh kg⁻¹)	Energy density (wh dm⁻³)	Opearating voltage(V)
Li	5928	7989	2.96
Na	2466	2466	2.3
K	1187	1913	2.37
Mg	5238	9619	2.09
Zn	1218	6136	1.66
Al	5779	10347	2.71
Fe	1080	3244	1.28

Aerogels II: Preparation, Properties and Applications   Materials Research Forum LLC
Materials Research Foundations **97** (2021) 99-120   https://doi.org/10.21741/9781644901298-6

3.1   Carbon aerogel

The carbon materials prepared from both bio-sources and synthetic chemicals make a major contribution to energy storage as they are cost effective and easily accessible [17]. Moreover, these kinds of renewable carbon materials may be produced using facile methodologies of pyrolysis and hydrothermal carbonization. On the other hand a great deal of aerogel materials have been exploited for various application including thermal insulation, high metal-absorption, catalyst supports and energy storage [18]. In this view, carbon aerogels (CAs) has drawn a great attention for various application which comprises high specific surface region and ultrahigh conductivity. They may be considered as potential electrode materials in energy transformation and energy storage gadgets including lithium-ion batteries, lithium-sulfur (Li-S) batteries, and supercapacitors [19–21]. Lithium - ion battery has turn into probably the great significance of chargeable and rechargeable batteries which is available on the market at 20 years ago. The batteries have unique features of high superior cell voltage, energy intensive and much lower security concerns. In case of anode materials for such systems, cyclability and capability forbids its commercial utilisation [22,23]. For example $LiCoO_2$, $LiNiO_2$ $LiMn_2O_4$, or $LiFePO_4$. In comparison to the above stated materials, graphite comprises the excellent capacity (372 mAhg^{-1}) due to excellent electrical conductivity and large surface area (400 to 1100 m^{2}g^{-1}) through which it can also be used as the negative electrode for the Li-ion batteries [24,25]. The discharge capacity associated with the parameters of certain porous carbon will increase on optimizing the pore diameter and pore volume. Usually, carbon aerogel were synthesized by sol gel and pyrolyzing process under inert atmosphere (N_2 or CO_2). The constant temperature can be flow through the tubular furnace with the charge of 2°C /min at different temperatures (600°C to 1100°C). After reaching the heat to 300°C at 2°C/min, the reactor temperature will increase to 3°C i.e maximum temperature. Using diamond saw the carbon aerogel can be cut into disk shaped electrode at 1mm thickness and the electrodes were heated at 300°C for 2 hours. Likewise several methods have been adopted for the facile synthesis of carbon aerogel followed by electrochemical property were analyzed initially before its application in batteries. For example Mojtaba Mirzaeian et al. [23] examined carbon aerogel prepared through the carbonization process under inert atmosphere. He varied the molar ratio of the carbon aerogel through sol gel followed by pyrolysis process. To manipulate the porosity of carbon aerogel. In this examination they found that the amount mesopore increases while increase the temperature. Here they obtain a mesopores material of excellent surface area (980.6 m^{2}g^{-1}) and high porosity of 99.6%. Lin Zhu et al. [26] prepared the carbon aerogel modified separator for lithium sulphur batteries via hydrothermal treatment followed by freeze drying process as the precursors of sweet

Aerogels II: Preparation, Properties and Applications | Materials Research Forum LLC
Materials Research Foundations **97** (2021) 99-120 | https://doi.org/10.21741/9781644901298-6

potato. The carbon aerogel modified separator from the lithium sulphur batteries exhibits the excellent electrochemical performances such as high discharge capacity of (1216 mAhg^{-1} and the capacity retention 40 % reversible discharge capacity over 1000 cycles (431 mAhg^{-1}). Fig. 2 [26] shows the schematic representation of the cell with the SP–CA modified separator.

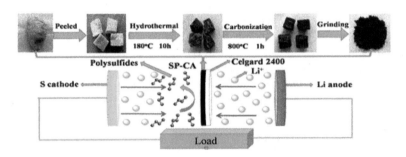

Fig. 2 *Schematic representation of the cell with the SP–CA modified separator [26].*

3.2   Graphene Aerogel

Nanotechnology exploits their significant related to the physical and chemical properties in nanoscale elements from zero (0D) to 3 dimension (3D). From the dimension 2D graphene aerogel has a excellent electrochemical performance and it is used in the energy storage application [27–29]. Presently, the balanced gathering associated with nano-sheets into macroscopic structures which have direct molecular and high internal reactive area are turn out to be used for energy storage and conversion. The 3D graphene aerogel have been used for the energy storage of the portable electronic devices which have exceptional high strength and mechanical stability. They have small meso and macro scales when compared to the 2D. The graphene aerogel has a large surface area (~2800 m^{2}g^{-1}), superior electrical conductivity (~10^{6}S cm^{-1}), less optical adsorption (2.2%) and thermal conductivity (~6000w/m/k). For instance Hui Shal et al. [30] have been studied the 3D graphene (GA) anode materials with surface defects through the facile hydrothermal procedure using graphene oxide. The three dimensional network material increased the electrical conductivity and density. These properties improve the electrochemical performance of energy storage made by anode materials. Lithium ion batteries are the outstanding energy storage devices compared with other batteries. Similarly Mercus A Worsely et al. [31] synthesized the graphene oxide based graphene aerogel with the high temperature up to 2500 °C. In these examination, graphene aerogel exhibit the pore volume (4 cm^3g^{-1} at 1500°C) and (2cm^3g^{-1} at 2500°C) and surface area (1200 m^2g^{-1}) at 1500°C) and (345 m^2g^{-1} at 2500°C). The pore volume and surface are decreases while increasing the temperature. But the pore size (>20 nm) increases while increasing the temperature and also increases the electrical conductivity by that order of magnitude. The Lin Liu et al. [32] prepared the graphene aerogel (GA) sponge by the frozen spray-coating technique. By varying experimental parameters the range of porosity of active material was modified as shown in fig. 3 [32]. The optimized porous material yields a large capacitance (230 F/g around 5000 cycles), better electromagnetic wave absorption capability and effective electrical loss. From these excellent properties, the GA can be applied to supercapacitors, battery electrodes, pollution absorption, electromagnetic shielding and so on [33,34]

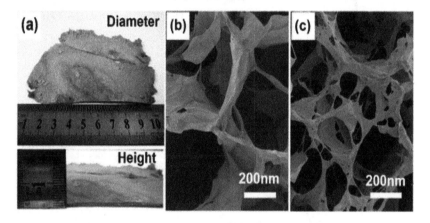

Fig. 3 *SEM micrograph of graphene aerogel (GA) sponge [32].*

3.3   Silicon aerogel

The silicon aerogel was first observed by Samuvel Stephes Kistle in 1930. He prepared silicon aerogel under inert atmosphere from ferric oxide, tin oxide, alumina, tungsten, cellulose, agar and egg whites [35–37]. Naturally the silica aerogel and synthetic silica has a foam like and amorphous structure. The silica has been made of sand, quartz and glass and mostly silica aerogel can be prepared from industrial waste as well as agriculture waste such as rice husk ash, rice hull, gold mine waste, coal gangue and so on. The agricultural waste was usually processed by any of the following three methods such as chemical, mechanical and enzymatic routes. The detailed processing procedures are shown in fig. 4 [38]. The rice husk ash exhibit 90 – 98 % of pure silica [39]. The silica

Materials Research Forum LLC
https://doi.org/10.21741/9781644901298-6

aerogel has a excellent energy storage used as a electronic devices compared with other aerogels. The aerogel exhibits the outstanding properties such as high surface area (100 m^2g^{-1}), low thermal conductivity (~0.001 $wm^{-1}k$), high porosity (99%) with the high optical transmission of 99% [38,40] The silica aerogel was used high temperature batteries through the excellent thermal properties.

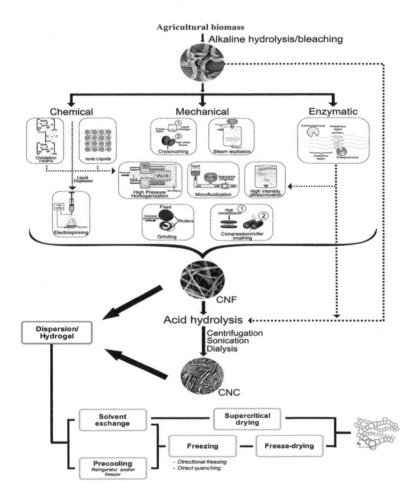

Fig. 4 *Different synthesis process of aerogel [38].*

For example Jiangxvan Song et al. [41] prepared the gel polymers silicon anodes for high performance battery by thermal crosslinking technique as the precursors of poly acrylic acid (PAA) and polyvinyl alcohol (PVA). At the current density of 4 Ag⁻¹, the prepared sample exhibits the high cycling stability of 1663 mAhg⁻¹ and excellent coulombic efficiency (99%). Similarly Manish Phadatane et al. [42] have studied the silicon nanoparticles from the silicon nanographite aerogel by aerogel fabrication method as the precursor of PVA. At the current density of 100 mA/g, the electrode have the specific capacity (455 mAh/g) and 97 % of coulombic efficiency with the good cyclic stability. During these properties, it has an excellent electrochemical performance of Li-ion batteries.

3.4   Metallic aerogel

The aerogel with enormous porosity, high surface area (400 – 1000 m^2g^{-1}) and low densities combine the physical and chemical properties of solid materials. The aerogel can be used in various applications such as batteries, supercapacitors and sensors. The aerogel include organic, inorganic and organic/inorganic composite [43]. Inorganic aerogels have been developed in more recent days. They include tellurides, metal alkoxides and selenides was discovered by Kanatzids, Brock and Co-researchers [44]. The metallic aerogel from the inorganic material was made by Leventis using elements like nickel, cobalt, cadmium, iron and copper. The metallic aerogel are synthesized by the sol gel method and it shown in the fig. 5 [45].

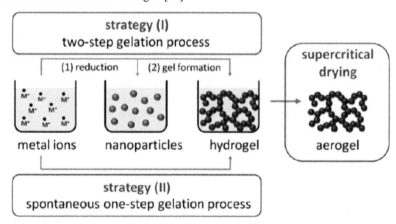

Fig. 5 *Synthesis process of metallic aerogel [45].*

For example Yaping P. Goa et al. [46] prepared the monolithic zinc oxide aerogel by a sol gel synthesis. The materials are annealing at lower temperature of below 250°C. In at low temperature the material exhibits high porosity and large surface area of 100 m^2g^{-1} and low density (~0.04gcm^{-3}). Similarly, Ryan P. Maloney et al. [47] prepared the lithium titanate aerogel for lithium ion batteries by the sol gel synthesis followed by freeze drying process. During calcination the lithium can vanish which increases the surface area. From the electrochemical analysis, in addition to redox peaks of commercial LTO electrode, the aerogel type LTO electrode exhibits an addition pseudocapacitive region due to its porosity and high surface area (30 m^2g^{-1}). This combined attributes of battery and supercapacitor in a same electrode may enhance the efficiency of the device. Hence such electrodes possess better electrochemical properties due to redox behavior.

3.5 Composite electrodes

Although the nano porous aerogel of several materials like silicon, carbon, graphene and so on possess large internal space with higher surface area it still does not meet the requirement of the real-world application. Again, to enhance the electrochemical attributes of aerogel electrodes several parameters like specific surface area, electrical conductivity, pore range and composition with other elements should be optimized. Subsequently, composition with other materials either metal or non-metal could further improve power density and cycle lifetime via synergistic effect. Unlike other materials whose reaction takes place on the surface aerogel as a host material facilitate facile penetration and diffusion of electrolyte ions deep in to the solid phase of active material thus paves to higher electrochemical performance [48].

As discussed Ronghua Wang et al. [49], prepared a 3D Fe_2O_3 nanocubes and loaded them in a N-doped graphene aerogel via a solvothermal induced self-assembly method. Optimized material from the work exhibits an enhanced performance in Li-storage attributes. The device exhibits a discharge capacity of 1774.6mAh/g that is comparatively better with its theoretical value (922.6 mAh/g). This may be attributed to a high active surface for electrochemical interaction and partial reversible process of solid electrolyte interface layer (SEI). It also holds a good cycling stability 108% of its initial capacity over 500 cycles with 100% coulombic efficiency. On the other hand long Liu et al.[50], synthesized Fe_2O_3 hollow sphere anchored N-Doped graphene aerogel through the Kirkendall effected facile process for the same lithium ion storage. In this case the Fe_2O_3 based graphene hybrid showed a high rate performance. Initial discharge capacity of about 902 mAh/g. Hence a rational design and a precise modification of morphology in aerogel and composite could modify the efficiency prominently. Other than this, metal oxides like cobalt, nickel, copper and so on can also be loaded on aerogel for its better application. Xin Sun et al. [51], prepared a $CoFe_2O_4$/ carbon nanotube (CNT) aerogel for li-ion batteries. The CNT and metal ion embedded in carbon aerogel yields a high specific surface area, withstands volume expansion and give a facile pathways for charge carriers. The optimized composite of materials gives outstanding reversible capacitance of 1033 mAh/g and approximately 80% of its capacity after 160cycles. Likewise Ping Wu et al. [52] made a composite aerogel comprising N-doped graphene and cobalt nickel sulphide through sulfur melting diffusion method for its application in Li-S batteries. As a conductive framework graphene facilitates an enhanced electrical conductivity. The composite shows discharge capacity around 1430 mAh/g. As a cathode of lithium sulphur battery the cathode exhibited its significant performance. The porous carbon /MoS_2 aerogel (VA – C/ MoS_2) was prepared by Peng Zhang et al. [53] through the hydrothermal process. In VA – C/ MoS_2 material its transfer resistance for ions has been decreased and also the transportation of ions and electrolyte are faster due to the vertically aligned pores. From the morphological characteristics, it provides good discharge capacity (1089 mAh/g) and extended cycle life nearly 2000 cycles. The areal capacity of it ranges 12.4 mAh/ cm^2 is 10 times increased when compared to the pristine MoS_2 aerogel. The advantages of morphology and outstanding redox behaviour of it has been used for the energy storage application of lithium-ion batteries. Fig. 6 [53] shows various steps involved in the fabrication of this composite aerogel with MoS_2 which uniformly clings on the lined up pores.

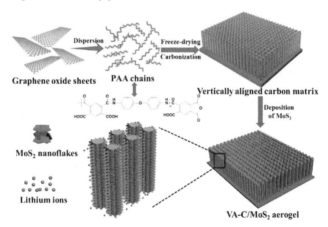

Fig. 6 *Schematic representation of fabrication procedure of VA-C/MoS₂ aerogel with MoS₂ [53].*

Lixiao et al. [54] prepared the three dimensional (3D) graphene aerogel with Fe_2O_3 through the hydrothermal treatment and followed by freeze drying process. The 3D Fe_2O_3/GAs have an excellent specific capacities (995 mAh/g) and rate capabilities (372 mAh/g) than the Fe_2O_3/GNs and it is applied for the energy storage application in the LIBS anodes. The author claims the high performance of the prepared Fe_2O_3/GAs was due to the Fe_2O_3 with the graphene aerogel of 3 dimensional structure. This yields a high surface area for the electrochemical reaction. For example Xi Wang et.al [55] describes the nitrogen doped G– SnO_2back to back aligned samples was produced via a new method. The structures can be formed by the 8-tetracyanoquinodimethane anion plays an important key role in the formation. When used as lithium-ion batteries, this material exhibits a high electrochemical performance, has a very large capacity, excellent cyclic stability and high rate capability. Yang Xie et al. [56] was prepared 3D boron-doped graphene aerogel (BGA) as cathode lithium sulphur batteries through hydrothermal method and it was filled by 59 % sulfur by which it can act as positive electrode in Li-S batteries which gives a enhanced capacity of 994 mA h $g^{-1}$and high rate capability compared with N-doped and pure graphene aerogel. In this case the capacity reaches nearly 600 mAh g^{-1}so it could be a potential electrode for Li-S batteries. Mohammad Akbari Garakani et al. [57] have been described the nanocomposite of ultra-ratio, cobalt carbonate nanoneedles and 3D porous graphene aerogel (CoCO_3/GA)prepared via hydrothermal method followed by freeze drying. After heat treatment, obtained the cobalt carbonate /GA and cobalt oxide (Co_3O_4)/GA anodes were used to the lithium-ion batteries. As resulting it has a good cyclic performance. The cyclic performances of cobalt oxide/GA electrode were slightly lower than the cobalt carbonate/GA electrode. So, the Co_3O_4/GA has a low lithium-ion storage capacity than the $CoCO_3$electrode. Both, the electrode had a high coulombic efficiency at a low current density of 0.1 Ag^{-1}. The electrochemical behaviour of the prepared sample exhibits the best performance as anodes in batteries. Jiarui He et al. [58] prepared a 3D CNT/graphene and LiS_2 composite aerogel and utilized it as cathode for lithium sulphur batteries. The constituents such as 1D CNT and 2D graphene nanosheet in composite provides a high mean free path to the transportation of charged particles and reduces the solubility ratio of polysulfides. The free standing architecture of electrode with porosity as seen in fig. 7 [58] records a reversible capacitance of 1123.6mAh/g.

Materials Research Forum LLC

https://doi.org/10.21741/9781644901298-6

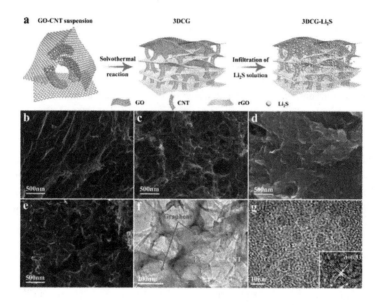

Fig. 7*(a) Schematic representation of preparation method of the 3DCG–Li$_2$S nanocomposite. scanning electron micrograph 3DG, 3DCG, 3DG–Li$_2$S, and 3DCG–Li$_2$S nanocomposite (b-e). TEM images at (f) Low-magnification of 3DCG–Li$_2$S, (g) Li$_2$S nanoparticles on 3DCG and its HRTEM image (Inset) [58].*

Ye Wang et al. [59] have prepared the 3D hybrid WS$_2$/carbon nanotube-rGO aerogels via hydrothermal process followed by freeze-drying. The 3D microchannel structure provides the excellent ionic conductive and also electronic transportation, hence leading to enhanced electrochemical properties for LIBs and SIBs. The specific capacities of lithium-ion batteries and the SIB's of product is 749 mA h g^{-1}. In comparison to noble and transition metals, earth abundant elements like Si, Sn, Ge and its relevant compounds holds good reversibility, which now-a-days are considered as potential material especially for commercial Li-ion based energy system. Hence, in comparison to single element in aerogel, the composite aerogel yields a better electrochemical behaviour.

Conclusions

In summary, aerogel based composites have exhibited a potential enhancement in both anode and cathode of different types of batteries. This could be due to the synergistic effect between the constituents of aerogel. However, in particular recent studies on ternary nanocomposites reveal the importance of two or three components excluding the drawback of single material aerogel. The usage of two or three elements as composites enhances electrical conductivity, electroactive site, energy density and stable morphology. On the other hand, it tailors the performance of batteries. In spite, high performance of the electrode materials of batteries still lag in efficiency of device performance for practical applications. There is still room for enhancement of performance. The major perspectives include (i) search of suitable material depending on battery (ii) designing of optimized morphology that facilitates a facile ion transportation channels, (iii) practicing an eco-friendly way of device fabrication& (iv) the mechanism involved in charge storage process and its mobility should be clearly investigated.

References

[1]   V. Etacheri, R. Marom, R. Elazari, G. Salitra, D. Aurbach, Challenges in the development of advanced Li-ion batteries: A review, Energy Environ. Sci. 4 (2011) 3243–3262. https://doi.org/10.1039/c1ee01598b

[2]   V. Thangadurai, S. Narayanan, D. Pinzaru, Garnet-type solid-state fast Li ion conductors for Li batteries: Critical review, Chem. Soc. Rev. 43 (2014) 4714–4727. https://doi.org/10.1039/c4cs00020j

[3]   L. Chen, L.Z. Fan, Dendrite-free Li metal deposition in all-solid-state lithium sulfur batteries with polymer-in-salt polysiloxane electrolyte, Energy Storage Mater. 15 (2018) 37–45. https://doi.org/10.1016/j.ensm.2018.03.015

[4]   Y. Li, H. Xu, P.H. Chien, N. Wu, S. Xin, L. Xue, K. Park, Y.Y. Hu, J.B. Goodenough, A perovskite electrolyte that is stable in moist air for lithium-ion batteries, Angew. Chemie - Int. Ed. 57 (2018) 8587–8591. https://doi.org/10.1002/anie.201804114

[5]   C.Z. Zhao, X.Q. Zhang, X.B. Cheng, R. Zhang, R. Xu, P.Y. Chen, H.J. Peng, J.Q. Huang, Q. Zhang, An anion-immobilized composite electrolyte for dendrite-free lithium metal anodes, Proc. Natl. Acad. Sci. U. S. A. 114 (2017) 11069–11074. https://doi.org/10.1073/pnas.1708489114

[6]   M.S. Whittingham, Lithium batteries and cathode materials, Chem. Rev. 104 (2004) 4271–4301. https://doi.org/10.1021/cr020731c

[7]   L. Ji, Z. Lin, M. Alcoutlabi, X. Zhang, Recent developments in nanostructured anode materials for rechargeable lithium-ion batteries, Energy Environ. Sci. 4 (2011) 2682–2689. https://doi.org/10.1039/c0ee00699h

[8]   B. Scrosati, J. Garche, Lithium batteries: Status, prospects and future, J. Power Sources. 195 (2010) 2419–2430. https://doi.org/10.1016/j.jpowsour.2009.11.048

[9]   X.B. Cheng, C. Yan, X.Q. Zhang, H. Liu, Q. Zhang, Electronic and ionic channels in working interfaces of lithium metal anodes, ACS Energy Lett. 3 (2018) 1564–1570. https://doi.org/10.1021/acsenergylett.8b00526

[10]   S. Wang, F. Gao, Y. Zhao, N. Liu, T. Tan, X. Wang, Two-dimensional CeO_2/RGO composite-modified separator for lithium/sulfur batteries, Nanoscale Res. Lett. 13 (2018). https://doi.org/10.1186/s11671-018-2798-5

[11]   P.G. Bruce, S.A. Freunberger, L.J. Hardwick, J.M. Tarascon, $LigO_2$ and LigS batteries with high energy storage, Nat. Mater. 11 (2012) 19–29. https://doi.org/10.1038/nmat3191

[12]   M. Winter, B. Barnett, K. Xu, Before Li ion batteries, Chem. Rev. 118 (2018) 11433–11456. https://doi.org/10.1021/acs.chemrev.8b00422

[13]   P. Albertus, S. Babinec, S. Litzelman, A. Newman, Status and challenges in enabling the lithium metal electrode for high-energy and low-cost rechargeable batteries, Nat. Energy. 3 (2018) 16–21. https://doi.org/10.1038/s41560-017-0047-2

[14]   G.J. May, A. Davidson, B. Monahov, Lead batteries for utility energy storage: A review, J. Energy Storage. 15 (2018) 145–157. https://doi.org/10.1016/j.est.2017.11.008

[15]   R. Pitchai, V. Thavasi, S.G. Mhaisalkar, S. Ramakrishna, Nanostructured cathode materials: A key for better performance in Li-ion batteries, J. Mater. Chem. 21 (2011) 11040–11051. https://doi.org/10.1039/c1jm10857c

[16]   C. Wang, Y. Yu, J. Niu, Y. Liu, D. Bridges, X. Liu, J. Pooran, Y. Zhang, A. Hu, Recent progress of metal-air batteries-A mini review, Appl. Sci. 9 (2019). https://doi.org/10.3390/app9142787

[17]   Y. Nishi, Performance of the first lithium ion battery and its process technology, Lithium Ion Batter. (2007) 181–198. https://doi.org/10.1002/9783527612000.ch8

[18]   Y. Gogotsi, P. Simon, Materials for electrochemical capacitors, Nat. Mater. 7 (2008) 845–854

[19]   L. Zuo, Y. Zhang, L. Zhang, Y.E. Miao, W. Fan, T. Liu, Polymer/carbon-based hybrid aerogels: preparation, properties and applications, 2015. https://doi.org/10.3390/ma8105343

[20]   J.C. Chang, Y.F. Tzeng, J.M. Chen, H.T. Chiu, C.Y. Lee, Carbon nanobeads as an anode material on high rate capability lithium ion batteries, Electrochim. Acta. 54 (2009) 7066–7070. https://doi.org/10.1016/j.electacta.2009.07.020

[21]   D. Guan, J. Shen, N. Liu, G. Wu, B. Zhou, Z. Zhang, X. Ni, The electrochemical performance of carbon-aerogel-based nanocomposite anodes compound with graphites for lithium-ion cells, J. Reinf. Plast. Compos. 30 (2011) 827–832. https://doi.org/10.1177/0731684411404458

[22]   S.J. Kim, S.W. Hwang, S.H. Hyun, Preparation of carbon aerogel electrodes for supercapacitor and their electrochemical characteristics, J. Mater. Sci. 40 (2005) 725–731. https://doi.org/10.1007/s10853-005-6313-x

[23]   M. Mirzaeian, P.J. Hall, Preparation of controlled porosity carbon aerogels for energy storage in rechargeable lithium oxygen batteries, Electrochim. Acta. 54 (2009) 7444–7451. https://doi.org/10.1016/j.electacta.2009.07.079

[24]   Y. Yan, M. Shi, Y. Wei, C. Zhao, L. Chen, C. Fan, R. Yang, Y. Xu, The hierarchical porous structure of carbon aerogels as matrix in cathode materials for Li-S batteries, J. Nanoparticle Res. 20 (2018) 1–13. https://doi.org/10.1007/s11051-018-4361-9

[25]   L. Yin, Z. Zhang, Z. Li, F. Hao, Q. Li, C. Wang, R. Fan, Y. Qi, Spinel $ZnMn_2O_4$ nanocrystal-anchored 3D hierarchical carbon aerogel hybrids as anode materials for lithium ion batteries, Adv. Funct. Mater. 24 (2014) 4176–4185. https://doi.org/10.1002/adfm.201400108

[26]   L. Zhu, L. You, P. Zhu, X. Shen, L. Yang, K. Xiao, High performance lithium-sulfur batteries with a sustainable and environmentally friendly carbon aerogel modified separator, ACS Sustain. Chem. Eng. 6 (2018) 248–257. https://doi.org/10.1021/acssuschemeng.7b02322

[27]   J.L. Brédas, E.H. Sargent, G.D. Scholes, Photovoltaic concepts inspired by coherence effects in photosynthetic systems, Nat. Mater. 16 (2016) 35–44. https://doi.org/10.1038/nmat4767

[28]   N. Bauer, K. Calvin, J. Emmerling, O. Fricko, S. Fujimori, J. Hilaire, J. Eom, V. Krey, E. Kriegler, I. Mouratiadou, H. Sytze de Boer, M. van den Berg, S. Carrara, V. Daioglou, L. Drouet, J.E. Edmonds, D. Gernaat, P. Havlik, N. Johnson, D. Klein, P.

Kyle, G. Marangoni, T. Masui, R.C. Pietzcker, M. Strubegger, M. Wise, K. Riahi, D.P. van Vuuren, Shared socio-economic pathways of the energy sector – quantifying the narratives, Glob. Environ. Chang. 42 (2017) 316–330. https://doi.org/10.1016/j.gloenvcha.2016.07.006

[29]   V.R. Stamenkovic, D. Strmcnik, P.P. Lopes, N.M. Markovic, Energy and fuels from electrochemical interfaces, Nat. Mater. 16 (2016) 57–69. https://doi.org/10.1038/nmat4738

[30]   H. Shan, D. Xiong, X. Li, Y. Sun, B. Yan, D. Li, S. Lawes, Y. Cui, X. Sun, Tailored lithium storage performance of graphene aerogel anodes with controlled surface defects for lithium-ion batteries, Appl. Surf. Sci. 364 (2016) 651–659. https://doi.org/10.1016/j.apsusc.2015.12.143

[31]   C. Zhu, T.Y.J. Han, E.B. Duoss, A.M. Golobic, J.D. Kuntz, C.M. Spadaccini, M.A. Worsley, Highly compressible 3D periodic graphene aerogel microlattices, Nat. Commun. 6 (2015) 1–8. https://doi.org/10.1038/ncomms7962

[32]   L. Liu, Z. Cai, S. Lin, X. Hu, Frozen spray-coating prepared graphene aerogel with enhanced mechanical, electrochemical, and electromagnetic performance for energy storage, ACS Appl. Nano Mater. 1 (2018) 4910–4917. https://doi.org/10.1021/acsanm.8b01091

[33]   J.Y. Huang, L. Zhong, C.M. Wang, J.P. Sullivan, W. Xu, L.Q. Zhang, S.X. Mao, N.S. Hudak, X.H. Liu, A. Subramanian, H. Fan, L. Qi, A. Kushima, J. Li, In situ observation of the electrochemical lithiation of a single SnO_2 nanowire electrode, Science (80-.). 330 (2010) 1515–1520. https://doi.org/10.1126/science.1195628

[34]   Y. Wang, H. Li, P. He, E. Hosono, H. Zhou, Nano active materials for lithium-ion batteries, Nanoscale. 2 (2010) 1294–1305. https://doi.org/10.1039/c0nr00068j

[35]   J. Song, M. Zhou, R. Yi, T. Xu, M.L. Gordin, D. Tang, Z. Yu, M. Regula, D. Wang, Interpenetrated gel polymer binder for high-performance silicon anodes in lithium-ion batteries, Adv. Funct. Mater. 24 (2014) 5904–5910. https://doi.org/10.1002/adfm.201401269

[36]   J. Song, Z. Yu, T. Xu, S. Chen, H. Sohn, M. Regula, D. Wang, Flexible freestanding sandwich-structured sulfur cathode with superior performance for lithium-sulfur batteries, J. Mater. Chem. A. 2 (2014) 8623–8627. https://doi.org/10.1039/c4ta00742e

[37]   B. Wicikowska, A.L. Oleksiak, silica aero-gel towards anodes for lithium-ion batteries, (2015) 9–10

[38]   J.L. Gurav, I.K. Jung, H.H. Park, E.S. Kang, D.Y. Nadargi, Silica aerogel: Synthesis and applications, J. Nanomater. 2010 (2010). https://doi.org/10.1155/2010/409310

[39]   N. Asim, M. Badiei, M.A. Alghoul, M. Mohammad, A. Fudholi, M. Akhtaruzzaman, N. Amin, K. Sopian, Biomass and industrial wastes as resource materials for aerogel preparation: opportunities, challenges, and research directions, Ind. Eng. Chem. Res. 58 (2019) 17621–17645. https://doi.org/10.1021/acs.iecr.9b02661

[40]   N. Recham, L. Dupont, M. Courty, K. Djellab, D. Larcher, M. Armand, J.M. Tarascon, Ionothermal synthesis of tailor-made LiFePO$_4$ powders for li-ion battery applications, Chem. Mater. 21 (2009) 1096–1107. https://doi.org/10.1021/cm803259x

[41]   T. Xu, J. Song, M.L. Gordin, H. Sohn, Z. Yu, S. Chen, D. Wang, Mesoporous carbon-carbon nanotube-sulfur composite microspheres for high-areal-capacity lithium-sulfur battery cathodes, ACS Appl. Mater. Interfaces. 5 (2013) 11355–11362. https://doi.org/10.1021/am4035784

[42]   M. Phadatare, R. Patil, N. Blomquist, S. Forsberg, J. Örtegren, M. Hummelgård, J. Meshram, G. Hernández, D. Brandell, K. Leifer, S.K.M. Sathyanath, H. Olin, Silicon-nanographite aerogel-based anodes for high performance lithium ion batteries, Sci. Rep. 9 (2019) 1–9. https://doi.org/10.1038/s41598-019-51087-y

[43]   N. Leventis, Three-dimensional core-shell superstructures: Mechanically strong aerogels, Acc. Chem. Res. 40 (2007) 874–884. https://doi.org/10.1021/ar600033s

[44]   W. Liu, A. Herrmann, N.C. Bigall, P. Rodriguez, D. Wen, M. Oezaslan, T.J. Schmidt, N. Gaponik, A. Eychmu, Noble metal aerogels- synthesis, characterization, and application as electrocatalysts, ACC. Chem. Res. (2014). https://doi.org/10.1021/ar500237c

[45]   H.D. Gesser, P.C. Goswami, Aerogels and related porous materials, Chem. Rev. 89 (1989) 765–788. https://doi.org/10.1021/cr00094a003

[46]   Y.P. Gao, C.N. Sisk, L.J. Hope-Weeks, A sol-gel route to synthesize monolithic zinc oxide aerogels, Chem. Mater. 19 (2007) 6007–6011. https://doi.org/10.1021/cm0718419

[47]   R.P. Maloney, H.J. Kim, J.S. Sakamoto, Lithium titanate aerogel for advanced lithium-ion batteries, ACS Appl. Mater. Interfaces. 4 (2012) 2318–2321. https://doi.org/10.1021/am3002742

Aerogels II: Preparation, Properties and Applications   Materials Research Forum LLC
Materials Research Foundations **97** (2021) 99-120   https://doi.org/10.21741/9781644901298-6

[48] J. Mao, J. Iocozzia, J. Huang, K. Meng, Y. Lai, Z. Lin, Graphene aerogels for efficient energy storage and conversion, Energy Environ. Sci. 11 (2018) 772–799. https://doi.org/10.1039/c7ee03031b

[49] R. Wang, C. Xu, J. Sun, L. Gao, Three-dimensional Fe_2O_3 nanocubes/nitrogen-doped graphene aerogels: Nucleation mechanism and lithium storage properties, Sci. Rep. 4 (2014) 1–7. https://doi.org/10.1038/srep07171

[50] L. Liu, X. Yang, C. Lv, A. Zhu, X. Zhu, S. Guo, C. Chen, D. Yang, Seaweed-derived route to Fe_2O_3 hollow nanoparticles/N-doped graphene aerogels with high lithium ion storage performance, ACS Appl. Mater. Interfaces. 8 (2016) 7047–7053. https://doi.org/10.1021/acsami.5b12427

[51] X. Sun, X. Zhu, X. Yang, J. Sun, Y. Xia, D. Yang, $CoFe_2O_4$ /carbon nanotube aerogels as high performance anodes for lithium ion batteries, Green Energy Environ. 2 (2017) 160–167. https://doi.org/10.1016/j.gee.2017.01.008

[52] P. Wu, H.Y. Hu, N. Xie, C. Wang, F. Wu, M. Pan, H.F. Li, X. Di Wang, Z. Zeng, S. Deng, G.P. Dai, A N-doped graphene-cobalt nickel sulfide aerogel as a sulfur host for lithium-sulfur batteries, RSC Adv. 9 (2019) 32247–32257. https://doi.org/10.1039/c9ra05202j

[53] P. Zhang, Y. Liu, Y. Yan, Y. Yu, Q. Wang, M. Liu, High areal capacitance for lithium ion storage achieved by a hierarchical carbon/MoS_2 aerogel with vertically aligned pores, ACS Appl. Energy Mater. 1 (2018) 4814–4823. https://doi.org/10.1021/acsaem.8b00897

[54] R. Wang, C. Xu, J. Sun, L. Gao, Three-dimensional Fe_2O_3 nanocubes/nitrogen-doped graphene aerogels: Nucleation mechanism and lithium storage properties, Sci. Rep. 4 (2014). https://doi.org/10.1038/srep07171

[55] X. Wang, X. Cao, L. Bourgeois, H. Guan, S. Chen, Y. Zhong, D.M. Tang, H. Li, T. Zhai, L. Li, Y. Bando, D. Golberg, N-doped graphene-SnO_2 sandwich paper for high-performance lithium-ion batteries, Adv. Funct. Mater. 22 (2012) 2682–2690. https://doi.org/10.1002/adfm.201103110

[56] Y. Xie, Z. Meng, T. Cai, W.Q. Han, Effect of boron-doping on the graphene aerogel used as cathode for the lithium-sulfur battery, ACS Appl. Mater. Interfaces. 7 (2015) 25202–25210. https://doi.org/10.1021/acsami.5b08129

[57] M.A. Garakani, S. Abouali, B. Zhang, C.A. Takagi, Z.L. Xu, J.Q. Huang, J. Huang, J.K. Kim, Cobalt carbonate/ and cobalt oxide/graphene aerogel composite

anodes for high performance li-ion batteries, ACS Appl. Mater. Interfaces. 6 (2014) 18971–18980. https://doi.org/10.1021/am504851s

[58]   J. He, Y. Chen, W. Lv, K. Wen, C. Xu, W. Zhang, W. Qin, W. He, Three-dimensional CNT/graphene-Li$_2$S aerogel as freestanding cathode for high-performance Li-S batteries, ACS Energy Lett. 1 (2016) 820–826. https://doi.org/10.1021/acsenergylett.6b00272

[59]   Y. Wang, D. Kong, W. Shi, B. Liu, G.J. Sim, Q. Ge, H.Y. Yang, Ice Templated free-standing hierarchically WS$_2$/CNT-rGO aerogel for high-performance rechargeable lithium and sodium ion batteries, Adv. Energy Mater. 6 (2016). https://doi.org/10.1002/aenm.201601057

Materials Research Forum LLC
https://doi.org/10.21741/9781644901298-7

Chapter 7

Aerogels Materials for Applications in Thermal Energy Storage

Sapna Nehra[1], Rekha Sharma[2], Dinesh Kumar[3,*]

[1]Department of Chemistry, Dr. K. N. Modi University, Newai, Rajasthan 304021, India

[2]Department of Chemistry, Banasthali Vidyapith, Banasthali, Rajasthan 304022, India

[3*]School of Chemical Sciences, Central University of Gujarat, Gandhinagar, India

* dinesh.kumar@cug.ac.in

Abstract

Over the years, aerogel materials reduced thermal conductivity, so proved to be the key method for preventing large consumption of thermal energy. In the class of insulating materials, aerogels have been found, these materials reduce the intermorphosis of heat between ambient sol−gel and various drying methods. Due to Aerogel's tremendous qualities, researchers and engineers showed keen interest in its construction. It showed various characteristics such as nano dimensions, minimum density, narrow, structured, small zero and exposed pore structure, forming through sol components in an arbitrary three-dimensional network. Notable, related to aerogel components, involves storage due to the significant capacity of thermal insulation and its minimum power of operation which means that heat can be stored for a longer period. Due to narrow structural entities, it easily captures light in the meso and nanoporous structure. Aerogels have a greater tendency regarding its heat storing efficacy, creating a simple nature, working consistency other than a commercial insulator. Therefore, this chapter focuses on aerogel's new strategy, which is constantly trending to increase the efficiency of aerogels and improving diverse structurally designed openings, especially insulation effectiveness and low thermal conductivity. Herein, we reviewed the formation of porous aerogels by using carbon nanomaterials, and their corresponding materials comprise GO, rGO, and fabrication with polymer, biomaterial which intrinsically embedded in the aerogel structure to achieve outstanding thermal storage characteristics for higher thermal behavior.

Aerogels II: Preparation, Properties and Applications
Materials Research Foundations 97 (2021) 121-144

Materials Research Forum LLC
https://doi.org/10.21741/9781644901298-7

Keywords

Carbon Nanotube, Reduced Graphene Oxide, Polymerization, Thermal Conductivity, Thermochemical, Composite

Contents

1.   Introduction

With the growing global population creating various crises such as water scarcity, polluted water, air pollution, energy storage, erosion in the abundance of fossil fuels, etc. In all these crises, access to energy sources is also limited, hence, attracting the attention of researchers [1-4]. Therefore, to overcome the energy crisis it is necessary to develop energy storage materials [5,6]. A variety of materials have been used in the storage of thermal energy, including organic and inorganic phase materials known as phase change materials (PCM). Further, inorganic PCM contains molten salt, hydrate salt, aqueous solution, etc. [7]. Many benefits found with inorganic PCM such as more latent heat production, lower and better thermal conductivity. However, some have been demonstrated problems such as phase separation, low stability, and supercooling. Therefore, it is very difficult to impregnate owing to their high polarity and water solubility [8]. Alternatively, organic phase change materials (OPCM) comprises polyethylene glycol (PEG), stearic acid, paraffin and fatty acids, etc. [9-12]. These materials are very outstanding because of their superb capability to absorb and evolved the latent heat at the time of phase transference [13-17]. All benefits, especially chemical stability, suitable phase transition temperature range, slight change in volume, attributed to the ability of organic phase materials in waste thermal solar and electrical energy applications [18-23]. Despite the excellent activity of organic phase materials, some characteristic shortcomings bound its utilization in various areas, for instance, lesser, thermal conduction, minimum energy conversion tendency, and outflow throughout melting and cooling procedures [24−26]. So, it heightens the necessity towards the generation of stable phase conversion materials with higher capabilities of energy alteration and heat conduction. Several tactics were employed to overcome these limitations like the use of polymer containing materials like in the form of polyurethane (PU), Polyethylene (PE), and Urea formaldehyde (UF) to maintain the organic phase material with an impregnated framework [27−29]. With the utilization of organic polymer, one issue of leakage has been resolved because of the flammable, less thermally conductive, and stable behavior. So, inorganic salt-based core- shell-like material as, silica oxide, calcium carbonate, and titanium oxide are utilized [30−32]. Organic polymers on relating to the inorganic shell materials exhibited the outstanding thermal conductivity, chemical stability, and protection. Meanwhile, because of many inorganic shells found in the composite resulted from an enhancement in the weight, which consequently reduced the latent heat [33]. Infusing the organic phase change material into spongy bulk materials, like metal−organic framework (MOF) diatomite, and perlite are very effective materials [34, 35, 24]. The porous materials show better results because of a porous network and big pore volume, which can easily trap the PCMs and release more latent heat [36]. But all these incorporated materials have some drawbacks as more weight and minimum chemical stability [37]. Here, 3-D carbon aerogels (CAs) nanostructures possess a porous network which attracts more attention because of their excellent features like higher thermal conductivity ultralight property and chemical stability [38]. Meanwhile, carbon materials show promising behavior in energy storage, high solar light absorption, and intertransformation of solar energy as desired [39]. For example, Zamiri et al. [40] mixed the 4% graphene into the phase change materials and gained the 150% increment in the heat conductivity of the developed composite. Characteristically, carbon aerogels CAs were prepared through the pyrolysis of resorcinol− formaldehyde (RF) grounded organic aerogel [41]. Both the carbon nanotubes (CNTs) and graphene have excellent nanostructures owing to their characteristic features displayed extreme fast growth in the arena of carbon aerogel [42]. However, the preparation of carbon aerogels includes complex preparation methods and exclusive and destructive precursors, resulting in higher costs and hampering large-scale application and meanwhile marking the load on the environment [43]. Therefore, it would be beneficial to develop highly efficient, environmentally friendly, cost-effective with high productivity nanocarbon aerogel [44].

Aerogels are known as open-cell materials that have very variable characteristics such as minimum density, high surface factor, maximum porous, minuet thermal conductivity low dielectric constant value, and many others [45,46]. In the year 1932, firstly, report the aerogel by Kistler obtained a wet silica gel without the destruction of the gel network through the supercritical drying [47−49]. Other than the silica, Kistler developed the various aerogels by using the different materials as tungstic, cellulose, ferric, alumina, stannic oxide, agar, gelatin, nitrocellulose, egg albumin nickel tartrate [50]. RF aerogel was developed in the 1980s by the Pekala using a polycondensation strategy. CAs were the product of pyrolyzing, which was generated in the 1990s [51,52]. But in recent times, besides RF and CAs several other fascinating materials are trending like CNT, graphene, and so on [53−58]. On behalf of the literature survey determined the formation of thermal insulation aerogel increased in recent years. By using an efficient sol-gel method developed, the varieties of aerogels comprise the xerogel, cryogel, and aerogel at the low-temperature condition. Aerogels are formed at that condition when the wet gel attained the critical temperature and pressure point in a given vessel, consequently some bind solvent release from the designed framework, then the formation of porous aerogel occurred. Xerogels formation is assigned to the evaporation of the liquid at ambient pressure and sublimation of freezing solvent corresponding to the cryogenics. In present time aerogels considered as minimum density with high surface area in which large-volume covered by pores means greater than 90% [59,60]. In the formation of different gels, a specific solvent required in the drying step means acetone or ethanol for both xerogel and aerogel and water for cryogel. Ultimately, current aerogels behave as accomplished materials owing to their excellent properties in many applications like auditory and thermal insulators, [61−63] absorption and desorption filters, sensors, electronics, aerospace, optical uses, energy storage devices automobile catalyst and biomaterials [64−73]. Mainly aerogels are classified into three main categories as inorganic, organic, and carbon. Metals like chromium, iron, aluminum, zirconium, and nonmetals like clay silica are used to coat the aerogels [74]. Silica categorized aerogels are consisting the silica oxide units and exhibited the average pore diameter in the between the dimensions of 10 to 100 nm and high porosity, high surface area, low bulk density from 85 to 99.8%, 600 to 1000 m^2/g, and 0.003 to 0.35 g/cm^3 respectively [75]. Usually, for the development of the aerogels, only two steps are essential one is gel formation and drying, while the next one step called the aging is necessary for strengthening the gel which obtained as a resultant of gelation [76]. However, the prominent thermal, physical, acoustical, and optical characteristics of silica aerogels are indisputable, but it exhibits the intrinsically hydrophilic behavior and delicate, which attributed the weak mechanical feature and limited their uses [77]. Other than these organic aerogels, such as related to the polymer, aerogels comprise the polyurea, polyimide, melamine-formaldehyde and PVA, RF [78,79] and cellulose, starch, and pectin relevant to the bio-based aerogels have been synthesized to deal with the limitation of the silica made up aerogels. Suitable features of organically related aerogels, like substantial mechanical nature and the less thermal conductivity, show their wide implication in the different fields [80].

Aerogels II: Preparation, Properties and Applications   Materials Research Forum LLC
Materials Research Foundations **97** (2021) 121-144   https://doi.org/10.21741/9781644901298-7

2.   Current status of the aerogel in thermal storage

Niu et al. [81] showed nanocellulose carbon aerogels aimed at storing thermal and derived energy with the help of using phase change material i.e. organic phase material (OPCM). It is a perfectly reasonable way to convert energy and store thermal and sunlight energy due to maximum porosity, high surface area minimum density, and maximum thermal conductivity of CA. The organic phase change material was allowed to be absorbed in high range due to their strong capillary behavior of porous nanofibers CA, which attributes about 96.2% energy change efficacy to stable size, large thermal conductivity, maximum latent heat, and phase change material i.e. carbon nanofibers aerogel. Here, with the help of a combination of phase change material and carbon nanofiber aerogel by doping of nitrogen and phosphorus type semiconductor. The gathering of heat and storage factor of phase change material carbon nanofibrous aerogels composites attributed the heat founder for that device required thermoelectric energy gathering. Around 55 and 80 mV energy voltage and their corresponding energy power density 50 W/m^2 and 1.20 W/m^2 were demonstrated by the thermo- and sunlight-driven energy gathering method correspondingly. These conversions of light signify the piratical application of thermoelectric devices for the use of clean energy in the form of heat waste and solar energy.

Zhang et al. [82] demonstrated the copper nanowire aerogels (CuNWAs) as a phase change material in storage energy devices, which have a very low density. With this property, low thermal conductivity has also been reported, which makes it inefficient as PCMs. For the increment in thermal conductivity phase change material i.e. copper nanowires fabricated with the paraffin and demonstrated the approximately 30% increment and shows the better electrical conductivity σ with 2.0 wt% of a copper nanowire. The obtained composite follows the shape stabilization factor even after the 50 cycles. It completely shrinks even performing every thermal cycle and shows the stability of approximately 130% with an increment of 6.3 wt% of a copper nanowire. To conclude, the analysis of both prominent factors like k- and σ, interfacial bonding maters in between the filler-filler and filler-matrix via the pre-sintering of a copper nanowire. Mainly in the pre-sintering process attempts two steps, one is the formation of bonding in between the copper nanowire and copper nanowire (CuNW-CuNW) and around over 200% σ- increment. The second step is the reduced the strength of bonding between the copper nanowire and the paraffin, with approximately 4% of k-reduction. The weakening of bonding among the filler-matrix interfacial well illustrated by the nano indentation method and computational modeling. Method of pre-sintering attributes the shape stabilization of the contrary to thermal cycling. Basically, it concluded that before that, no metallic aerogel was used to impregnate the shape maintained electrically conductive thermal-energy storage materials (TESMs) having over 130% of k-increment. Here, obtained the pre-sintered copper nanowall paraffin composite showed the σ value is equal to the =14 S/m. In the present study found that the two dominant actors like σ- and k- increments through the filler-filler and filler-matrix interfacial bonding, respectively. This kind of prominent study gives a shred of strong evidence to design the high thermal energy storage materials having maximum electrical and thermal conductivities with maximum durability.

Fan et al. [83] reported the formation of graphene aerogel exhibited the highest thermal, mechanical, and electrical features by using three different reducing agents, including L-ascorbic acid, HI, and NaHSO3. They have studied very well the effect of used three different reducing agents on the shape and features of the developed aerogels. The effect of thermal annealing was also demonstrated in the presence of argon gas atmosphere at 450 ˚C. The applied thermal annealing process increases the electrical conductivity of the graphene aerogels up to five times and also affects the corresponding surface area. Particularly, regarding the different reducing agent and CNT concentration shows the different surface areas like around 30% reduced and 15% increased corresponding to the HI and NaHSO3 and LAA. Compare to the bare graphene aerogel, mixing of the CNT increases the surface area and electrical conductivity about 14% and 2.8 times, respectively. The infrared microscopy method was utilized to measure the thermal conductivity was approximately 0.10 W/(m K). The deep study and optimization of graphene aerogels attribute the wide application in different fields like the formation of an electrode, manufacturing of storage devices, and in the development of nanocomposites.

Mu et al. [84] reported the novel thermal energy storage as form-stable composite phase change materials (FS-CPCM) by the utilization of the vacuum fabrication process. During synthesis, lauric acid act as a phase change material and fabricated over the surface of GA through the esterification and a reduction method. It also works as supporting material and enhances the thermal conductivity of the nanocomposite. In the form-stable composite phase change materials, various other properties also were investigated, for instance, the structural behavior, thermal energy storage, and the conductivity by using the various sophisticated techniques respective to the FESEM, FTIR, and XRD. These techniques illustrated the fabrication of the lauric acid over the surface of graphene oxide and formed the LA-GA structure and found no disturbance in the chemical structure of the lauric acid during the vacuum impregnation. By the fabrication obtained two types of the nanocomposite, one was LA/LA-GA FS-CPCMs, and another was LA/GA FS-CPCM. Among both nanocomposites, a better crystallinity of lauric acid was found in the LA/GA FS-CPCM. The thermal conductivity of the lauric acid was increased by the addition of graphene and formed the lauric acid fabricated graphene aerogel. The thermal conductivity was observed in the LA/GA FS-CPCM around 1.207 W/m K, which was 32.6 and 352.1%

higher than the LA/GA FSCPCM and lauric acid, respectively. The LA/GA FS-CPCM exhibited the excellent potential of phase changing capability, minimum cooling temperature, and good thermal stability, and it attained the 207.3 J/g and 205.8 J/g, of melting and freezing enthalpy correspondingly. Other than these, it remained the superb light absorption and simultaneously conversion into the heat form and could convert the around 80.6% light into the heat form. These characteristics feature showed the LA/LA-GA FSCPCM's utilization storage of thermal energy storage and in the solar energy storage systems.

Hong et al. [85] attempted to overcome the crises of energy by preparing the super wetting aerogels for the energy polypropylene storage and the thermal energy storage for sustainable development. During the synthesis, they impregnate the polypropylene aerogels for the formation of phase change material composite as the support. Because of the various superb features of aerogels like higher porosity, lesser weight, excellent super-hydrophobic as well super oleophilic tendency, a lipophilic organic phase change material can be easily fabricated over the polypropylene (PP) aerogel. With maximum loading capacity around 1060 wt%, which was 2 times higher than the other PCMs composites. The polypropylene phase change materials exhibited the very good thermal stability and recyclability excellent latent energy up to 141.1 to 159.5 kJ kg^{-1}. These stabilities remained constant even after the 50 successive cycles of melting and freezing. XRD results reveal the loading of the organic phase change material into polypropylene aerogel network reduces the crystal size of their corresponding network of composites. The outstanding feature of the polypropylene phase change materials displays the increasing thermal conductivity, regarding the polypropylene fabricated over the paraffin composite, which was over two times of paraffin. As a result of the minimum cost of polypropylene, easy operation process, maximum material loading property, excellent stability, outstanding cycling behavior, and heat conductance attributed the huge potential to the polypropylene fabricated phase change material for renewable energy storage implications. Overall, it concluded that all unique properties of the super wetting feature of aerogels open up a new vision to create the high-performance phase change material to diminish the energy storage crises at a large scale.

Li et al. [86] prepared a promising composite for the wide application in energy storage. Because of characteristic properties including the minimum density, maximum electric conductance, porosity, chemical dullness, and so forth. Recently, researchers have been trying to focus on the diverse synthesis method for the synthesis of the RF-CAs like the drying, sintering temperature, polymerization, activation methods, and so on. But, the well-described information about the surface chemistry, microstructure, and their employment in energy storage devices of the resorcinol-formaldehyde based carbon aerogels are hardly discovered. Therefore, the present study examined the application of the RF-CAs in various areas of energy storage, including Li, Na, Li-S, ion batteries. Li and coworkers focus on illustrating the various structural behavior interlinks and their corresponding synthetic routes.

Shen et al. [87] examined PEG, which was not an efficient support phase-change material because of the low thermal conductivity, less stability, and outflow tendency. Therefore, used a new phosphorylated polyvinyl alcohol (PPVA) and GA were used as a support matrix for preparing the PPVA/GA/PEG PCM. Here, the PPVA and GA formed the double-network as the support material and synthesized by the single-step and eliminated the drawbacks of the single PEG. Around only 15% involvement of the all three PPVA, GA, and PEG as a phase change material in composite showed the maximum energy storage capacity, large temperature stability and exhibited the maximum shape-stabilization. Only 15% containing PPVA/GA/PEG composite exhibited the thermal conductance of 0.610 W m^{-1} K^{-1}, by loading of 1.60 wt% GA compared to polyethylene glycol was 0.493W m^{-1} K^{-1}. This PCM composite showed 119.6 J g^{-1} of the latent heat energy, which was in adequate. Around 19.2% of heat was reduced. The intensity in the temperature's peak elimination rate reduced by 19.2% regarding polyethylene glycol. Beside all properties, it shows the high anti-leakage property with the maximum temperature stability, and it can sustain their corresponding stability even after the 50 cycles. So above all discussed outstanding property illustrated the application in wide energy storage fields

Tang et al. [88] used both PEG and GA as the PCMs by given different oxidation levels to the graphene oxide during the Hummers method. Here, during synthesis by incorporating the polyethylene glycol into GAs through the vacuum fabrication method. Via the oxidation process of graphene oxide, the framework of the graphene aerogel can be mold. Several techniques were employed to determine the structure of the PEG- GA PCMs like XRD, FTIR, XPS, and Raman spectroscopy. These techniques assured the by enhancing the oxidation levels, the number of oxygen moiety eased, similarly OH groups interchange into the carboxyl and epoxy functional moiety. In the oxidation process, the graphitic nature of graphene oxide reduces, though the sp^3 hybridization of graphene oxide is enhanced because of the interruption in the sacking order of graphite. The surface morphology depicts the disruption in graphene sheets and seems to be more dangerous by raising the oxidation level. With the high oxidation process of graphene oxide formed graphene aerogels, resulted PCMs displayed the superb shape stable and outstanding thermal repeatability. In the seven's presence newtons of force, a negligible change in the structure of the PGA6-40 when the temperature was given from 35 to 150 °C. Further, PCMs show a very significant interconversion of the photo-to-thermal energy and keep the thermal energy very well Huang et al. [89] illustrated the phase transfer behavior of the carbon aerogels for thermal energy storage. Here, phase transfer phenomena octadecanoic impregnated with the glucose derivative carbon aerogel and observed the disturbance in the regular arrangement of the octadecanoic molecule through the amorphous CA. The molecular detachment increases splits solid-solid and solid-liquid phase change method. They have investigated the PCMs as an interconversion of a photo to thermal energy and further used for commercial application. The synthesized nanocomposite possessed maximum energy capacity compared to the bare PCMs, which can be attained a high temperature at 62.2 °C. This present study tested the different strategy to examine a phase change behavior of PCMs and behaves the reservoir of photon energy to the thermal energy.

Truong Nguyena et al. [90] developed the morphology-controlled graphene aerogel for thermal energy storage. In recent time over the GA attracts the many researchers in various energy applications owing to their outstanding properties like having lesser temperature resistivity, maximum surface area, mechanical flexibility, more carrier agility, and minimum cost, a simple approach in interconvertible of graphene to graphite. The working potential graphene aerogel-based electrode lies on the surface and their structure of graphene aerogel. Although in this study, some parameters were evaluated to optimize the graphene aerogel morphology and the structural basis like density, surface area, pore size distribution, and its volume. Here, the graphene aerogel structure was prepared using practically available graphite using the hummers method. Because of cost-effectiveness, easy operationality, and eco-friendly nature, showed the hydrothermal method was an efficient method and used for preparing graphene aerogel. There is no binder used as a support during the synthesis to restrict the adverse effect of the electrical conductance of GA. Impact of impregnating terms and conditions, the concentration of graphene oxide in the formation of nanostructural graphene aerogel was also measured. When the concentration of graphene oxide was 3 mg/mL utilized, then the BET surface area of 394 m^2/g with 0.042 g/cm^3 density was obtained after the hydrothermal treatment at 180 °C for 1.5 h. The temperature durability test examined the graphene aerogel can be stable up to 500 °C in the atmosphere till 24 h time period was given during hydrothermal treatment. Initially, the graphene aerogel shows the 0.004 S/cm of electrical conductance after 6 h hydrothermal treatment. All experimental and physiochemical studies confirmed the graphene aerogels have an application in energy storage in the thermal form.

Wang et al. [91] developed the flexible 3-D nanocomposite aerogel of boron nitride (BN) via the self-aggregation of the three-dimensional interconnected framework of inorganic and organic functional substituents. Both precursors, as functional boron nitride (FBN) and polyimide (PI), enhance the mechanical strength in the presence of compression, coagulation, or elongation of the aerogels because of their synergistic effect. The 3-D thermal conductors exhibited the count of high-density bulk solid, having weighty nature difficult in transporting manner to reach the gratify present demands in elastic and soft electronic devices. Therefore, to access their portability in the flexible devices, a lesser weight containing highly flexible 3-D aerogels was designed through a freeze-drying strategy. The obtained aerogel was establishing minimum bulk density around 6.5 mg cm^{-3}. Although, BN aerogels maintain the high thermal stability and conductivity even at 30-300 °C temperature range confirmed the use of this aerogel in enormous temperature containing environments. It can behave as minimum weight and flexible temperature conductive to increase the temperature energy storage. Remarkably, in the presence of an alternative stress cycle within the temperature and the high flexibility, randomness of heat conductance makes the aerogel allow to arrest the thermal energy.

Yang et al. [92] developed 3-D hybrid graphene aerogel (HGA) fabricated on graphene foam formed via the CVD method implied for commercial application. The attained composite PCMs with heated conductance and GA showed an effective support scaffold to expand the shape permanency of organic phase change materials. HGA was impregnated in a graphene oxide structure to attain a three-dimensional graphene from a hybrid graphene aerogel microstructure. With comparison to bare paraffin, wax and PW/GF thermal conductivity of paraffin wax (PW)/GH composite PCMs increases by 574% and 98% compared with pure PW and PW/GF composite PCMs, respectively. In the meantime, paraffin wax graphene aerogel phase change material composite shows the stable morphology with comparison to the paraffin wax graphene foam phase change material contained maximum temperature energy storage density, better thermal consistency, and organic stable. Paraffin wax graphene microstructure confirmed an efficient PCM for the conversion of natural light into thermal energy conversion and excellent restoration and photo-absorption. So, the current study opens up the way to synthesize the potential phase change material composite with characteristics features which remarkably used in the energy restoration.

Yang et al. [93] developed the polyethylene glycol-based phase change material composite exhibited the maximum stabilization, maximum thermal energy storage capacity, worthy conversion of the light into heat form. They have fabricated the PEG through the vacuum impregnation method into the HGA, which were the combination of the graphene oxide and the graphene nanoplate. As a resultant obtained the maximum energy storage, conductivity, stable morphology, and very renowned thermal cyclability, attributes the easy photon conversion to heat form. Here, graphene sheets generate the 3-D network to maintain the structure of the polyethylene glycol at the time of altering the phase. Graphene nanoplate homogeneously distributed over the wide network of the graphene oxide and formed a suitable conductive path regarding the thermal channel. The involvement of the hybrid graphene aerogel commendable improved the conductive behavior and attributes the stable morphology character to the phase change material composite. With polyethylene glycol hybrid graphene aerogels, phase change material containing the 0.45, 1.8 wt.% graphene oxide and graphene nanoplate, respectively. It possesses the enhanced thermal conductance of the 1.43 W/mK, which improved from the 0.31 W/mK of pure polyethylene glycol. Therefore, this appropriate phase change material best suitable for the energy conversion process from light to temperature stably. It proves as a simple and greener method to achieve the capable composite which exhibited the higher thermal conductivity and restore the energy density and molded the phase change material in easy interconversion of the form photon to heat energy.

Niu et al. [94] demonstrated the highly efficient organic phase change material to control the energy crisis. In front of researchers, a big challenge to conserve the sunlight energy into the useable form. Here synthesized the phase change material by involving the carbon nanofiber aerogel and contribute to the energy storage system. The obtained OPCM shows a higher surface area, minimum density volume, and higher thermal restoration capacity because of the contribution of the wood-derived carbon nanofiber. Mainly the capillary phenomena in the wood carbon nanofiber aerogel provided the absorption of OCPMS to a large extent and gave the sign of shape stability, maximum latent heat, and maximum heat capacity. They reported approximately 96.2% heat conversion in the CNFA composite and illustrated the insertion of N and P-type semiconductors during the fabrication of solar and light device in the form of PCM/CNFA. Both heat storage and thermoelectric effect simultaneously work and support each other. The dynamism collecting expedient shows a higher power density of 0.50 W/m^2 in the sunlight's interconversion derived thermoelectric devices was 1.20 W/m^2 that promises for the commercial growth of a thermoelectric system too applicable to sparkling energy like waste heat and solar energy.

Im et al. [95] illustrated the interchange of discarded heat energy into the electricity form and proved to the best strategy to resolve the energy crises facing the present world. They developed the electrochemical electrode by using efficient material like carbon nanotube and transformed into the aerogel form and used for the storage of heat energy. Although, the largest available thermal energy cannot be collected easily because of less availability of the efficient and economically viable system, which can inter-convert the discarded heat into the electricity. So, in this strategy, easily incorporated the CNTA related thermo-electrochemical cells, exhibited very less cost, maximum efficiency for the further implication in the thermal storage system. The highest power efficiency around 6.6 Wm^2 at 51°C with comparable Carnot efficacy of 3.95% was reported when standardized the electrochemical cell cross-sectional area. In the present method, the electrode found the

Materials Research Forum LLC
https://doi.org/10.21741/9781644901298-7

significance of purity, good porosity, and outstanding surface to enhance the activity of the thermo-cell in energy harvesting.

Zhao et al. [96] demonstrated the highly efficient PCCMs show the maximum latent heat efficiency, tough morphology, and possessed the maximum efficiency towards the thermal energy. They designed the phase change material by formation of the rGOABs and involvement of 1-tetradecanol (TD) paraffin, which acts as heat restoring material, and graphene oxide behaves as support material. The latent heat in the novel composite was 230.3 J g^{-1} estimated by differential scanning calorimetry. Even after fifty runs of heating and cooling effect rGOABs maintain its 96.6% efficacy promises the best thermal storing composite for their practical application. The structural units rGOAB/TD are almost similar to the individual 1-tetradecanol and graphene oxide means during the fabrication, no ambiguity comes in both structures. The obtained composite contains the 98.83 wt% of TD, which has ever been reported a large amount of TD in the composite. All thermal experiment results illustrated the outstanding thermal conductivity of the rGOAB/TD over the GOAB/ TD due to a sort of graphitic reduction. The rGOAB/TD PCCM composite shows the highest heat restoration capacity, good elasticity with the maximum transference of the heat. Further, its applications have a wide tendency in practical thermal applications like in Li-ion batteries.

Liu et al. [97] described the various targets by illustrating the PCMs having wide, capable forthcoming issues like a collection of the latent heat, interconversion of the sunlight so, further employed in the thermal applications. In promising novel work, they have developed the two different diameters containing rGO by the mixing of the two-phase change materials as octadecanoic and stearic acid correspondingly. Both organic phase material was incorporated in the 3D framework of the graphene oxide with the help of hydrothermal strategy. Few studies confirmed the reduced graphene oxide largely affects the latent heat, and their corresponding produced conductance of a diverse variety of organic phase change materials. With integrating 10 wt% reduced graphene oxide increased 3.21 $Wm^{-1}K^{-1}$ of thermal conductance. Subsequently, some graphene parts replaced by some silver nanoparticles. After the substitution with sliver nanoparticles achieved enhanced thermal conductivity as 5.89 $Wm^{-1} K^{-1}$. The reduced graphene oxide size remarkably affects the thermal contact confrontation. Basically, they found the synergism in between the reduced graphene oxide framework and the silver nanoparticle which profits conductance of newly formed nanofabricated PCM as NePCMs.

Wang et al. [98] demonstrated the graphene oxide aerogel as PCM for the thermal restoration, which used as a thermal shield in electronically devices i.e. electric circuit and microelectronic strategies. Through the ink-jetting method (ILS) described the homogenous distribution of GA microspheres with flawless spherical diameter. Via the ink-jet method referred, the aerogel microspheres might be measured through regulating the database of the ink-jetting method. When the temperature increased from 20, 25, 30 °C diminished the BET surface area 444.3, 330.9, 209.8 m^2/g, correspondingly. The optimized experimental studies estimated the electrical resistance, the water contact angle of graphene aerogel microsphere below than the 8.79°, and over 137.5°, respectively. Subsequently, homogenous GAM act as support, wherever vacant cites filled by the paraffin PCM. So, a few of PCM was obtained into thermal flux buffer. Because of the presence of micro and nanopore canals in graphene aerogel microsphere, the maximum crystalline, latent heat, and specific molded shape of paraffin could be well sustained and optimized. The resulting GAM shows steady thermal behavior at the time of the phase change method. Because of micron size, overall heat subsidized to a tiny extent in phase altering of paraffin.

Liang et al. [99] used the vacuum-assisted impregnation method followed by the reduction reaction for the synthesis of graphene-coated graphene material via used the freeze-drying method, subsequently shape stabilized phase change material grounded over the Lauric acid and graphene/GO complex aerogels were successfully fabricated. Various analytical tools were utilized to check the structural composition of the composite that was SEM, XRD, and FTIR. Not only lauric acid incorporated aerogel formed virtuous figure and heat stability meanwhile displayed the maximum thermal and electrical conductivity. The lauric acid aerogel composites attained maximum latent heat during the phase transference medium was 198 J/g because of the 97 wt% integration of the lauric acid over 97 wt%. So, it possesses the very good thermal restorative energy efficacy, maximum thermal switch on and off capacity, outstanding cyclical constancy, and best phase-transition reversibility. Kinetic study of crystallization revealed that the aerogel framework could produce the assorted nucleation influence over the crystallization of lauric acid. So, the speed of crystallization a lauric acid laden in the complex aerogels was remarkably increased in the isothermal's presence, and the non-isothermal method attributes the lauric acid rapid thermal active at even in room temperature condition.

Zhou et al. [100] utilized the eminent method like hydrothermal and froze drying for the formation of GOA alleviated by a diverse range of impregnation of the $MgCl_2 \cdot 6H_2O$ in the composite. With the help of differential scanning calorimetry (DSC) evaluated the influence of GOA over the dryness temperature of GOA- $MgCl_2 \cdot 6H_2O$ composites. Because of its nano range size, demonstrated the minimum crystallinity of magnesium chloride hexahydrate over the inward surface of graphene oxide aerogel well estimated by the XRD and TEM techniques. When increased the quantity of the graphene oxide aerogel, the chief dehydration peaks in the spectral change towards the lesser temperature. By loading 50 wt% magnesium chloride hexahydrates, it reduced at 90 °C in the fourth endothermic peak of temperature regarding the bare magnesium chloride hexahydrate.

Still, this is an only first-time reported composite which altered by the varying temperature condition of hydrate salt was demonstrated. The salt hydrated composite showed the 1598 J/g highest energy density. Therefore, graphene aerogel fabricated hydrated salt played a remarkable effect in lesser temperature thermochemical restoration.

Conclusions

The present chapter discusses the newly trending restorative energy material to synthesize energy materials. Aerogels are largely utilized materials for the sensible restoration of the thermal energy and assigned as cost-effective, maximum storage capacity, and eco-friendly material. Wide utilization of the various scientific strategies emphasize the use of aerogels to manage the energy storage system and fulfill the demand of energy. Other than the thermal energy, it used in the multidisciplinary areas like in the removal of toxic contaminants, environmental remediations, water purification. In multidisciplinary fields, researchers demonstrated several aspects and claimed the tendency of the aerogels. A specific chemical modification offers the compositional changes can be alternative media used in the multiple devices.

Acknowledgments

Dinesh Kumar is thankful DST, New Delhi, for financial support to this work (sanctioned vide project Sanction Order F. No. DST/TM/WTI/WIC/2K17/124(C).

References

[1] Midilli, M. Ay, I. Dincer, M. A. Rosen, On hydrogen and hydrogen energy strategies: I: current status and needs, Renew. Sustain. Energy Rev. 9 (2005) 255–271. https://doi.org/10.1016 rser.2004.05.003

[2] L. Yang, Z.G. Chen, M.S. Dargusch, J. Zou, High performance thermoelectric materials: progress and their applications, Adv. Energy Mater. 8 (2018) 1701797–1701797. https://doi.org/10.1002/aenm.201701797

[3] Y. Yang, W.X. Guo, K.C. Pradel, G. Zhu, Y. Zhou, Y. Zhang, Y. Hu, L. Lin, Z.L. Wang, Pyroelectric nanogenerators for harvesting thermoelectric energy, Nano Lett. 12 (2012) 2833–2838. https://doi.org/10.1021/nl3003039

[4] Y.M. Wang, B.T. Tang, S.F. Zhang, Single-walled carbon nanotube/phase change material composites: sunlight-driven, reversible, form-stable phase transitions for solar thermal energy storage, Adv. Funct. Mater. 23 (2013) 4354–4360. https://doi.org/10.1002/adfm.201203728

[5] T.Y. Kim, J. Kwak, B. Kim, Energy harvesting performance of hexagonal shaped thermoelectric generator for passenger vehicle applications: an experimental approach, Energy Convers. Manage. 160 (2018) 14−21. https://doi.org/10.1016/j.enconman.2018.01.032

[6] C.B. Vining, An inconvenient truth about thermoelectrics, Nat. Mater. 8 (2009) 83−85. https://doi.org/10.1038/nmat2361

[7] P. Zhang, X. Xiao, Z.W. Ma, A review of the composite phase change materials: fabrication, characterization, mathematical modeling and application to performance enhancement, Appl. Energy,165 (2016) 472−510. https://doi.org/10.1016/j.apenergy.2015.12.043

[8] Z.F. Liu, Z.H. Chen, F. Yu, Preparation and characterization of microencapsulated phase change materials containing inorganic hydrated salt with silica shell for thermal energy storage, Sol. Energy Mater. Sol. Cells, 200 (2019) 110004. https://doi.org/10.1016/j.solmat.2019.110004

[9] J. Wang, H. Xie, Z. Xin, Thermal properties of paraffin-based composites containing multi-walled carbon nanotubes, Thermochim. Acta 488 (2009) 39−42. https://doi.org/10.1016/ j.tca.2009.01.022

[10] Y. Jiang, E. Ding, G. Li, Study on transition characteristics of PEG/CDA solid-solid phase change materials, Polymer, 43 (2002) 117−122. https://doi.org/10.1016/S0032-3861(01)00613-9

[11] D. Feldman, M.M. Shapiro, P. Fazio, A heat storage module with a polymer structural matrix, Polym. Eng. Sci. 25 (1985) 406−411. https://doi.org/10.1002/pen.760250705

[12] A. Sari, C.Alkan, U. Kolemen, O. Uzun, S. Eudragit, Methyl mathacrylate methacrylic acid copolymer)/fatty acid blends as form stable phase change material for latent heat thermal energy storage, J. Appl. Polym. Sci. 101 (2006) 1402−1406. https://doi.org/10.1002/app.23478

[13] G.Y. Fang, H. Li, Z. Chen, X. Liu, Preparation and characterization of stearic acid/expanded graphite composites as thermal energy storage materials, Energy, 35 (2010) 4622−4626. https://doi.org/10.1016/j.energy.2010.09.046

[14] Y. Li, Y.A.S Amad, K. Polychronopoulou, S.M. Alhassan, K. Liao, From biomass to high performance solar-thermal and electric-thermal energy conversion and storage materials, J. Mater. Chem. A, 2 (2014) 7759−7765. https://doi.org/10.1039/C4TA00839A

[15] Y. Yoo, C. Martinez, J. Youngblood, Synthesis and characterization of microencapsulated phase change materials with poly (urea- urethane) shells containing cellulose nanocrystals, ACS Appl. Mater. Inter. 9 (2017) 31763−31776. https://doi.org/10.1021/acsami.7b06970

[16] A. Sharma, V.V. Tyagi, C.R. Chen, D. Buddhi, Review on thermal energy storage with phase change materials and applications, Renew. Sustain. Energy Rev. 13 (2009) 318−345. https://doi.org/10.1016/j.rser.2007.10.005

[17] J. Puig, I.E.dell' Erba, W.F. Schroeder, C.E. Hoppe, R.J.J. Williams, Epoxy-based organogels for thermally reversible light scattering films and form-stable phase change materials, ACS Appl. Mater. Inter. 9 (2017) 11126−11133. https://doi.org/10.1021/acsami.7b00086

18] F. Agyenim, N. Hewitt, P. Eames, M. Smyth, A review of materials, heat transfer and phase change problem formulation for latent heat thermal energy storage systems (LHTESS), Renew. Sustain. Energy Rev. 14 (2010) 615−628. https://doi.org/10.1016/j.rser.2009.10.015

[19] J.L. Yang, L.J. Yang, C. Xu, X.Z. Du, Experimental study on enhancement of thermal energy storage with phase-change material, Appl. Energy, 169 (2016) 164−176. https://doi.org/10.1016/j.apenergy.2016.02.028

[20] S. Wang, P. Qin, X. Fang, Z. Zhang, S. Wang, X. Liu, A novel sebacic acid/expanded graphite composite phase change material for solar thermal medium-temperature applications, Sol. Energy, 99 (2014) 283−290. https://doi.org/10.1016/j.solener.2013.11.018

[21] L.W Fan, X. Fang, X. Wang, Y. Zeng, Y.Q. Xiao, Z.T. Yu, X. Xu, Y.C. Hu, K.F. Cen, 2013. Effects of various carbon nanofillers on the thermal conductivity and energy storage properties of paraffin-based nanocomposite phase change materials. Appl. Energy, 110, pp.163-172. doi.org/10.1016/j.apenergy.2013.04.043

[22] L. Chen, R.Zou, W. Xia, Z. Liu, Y. Shang, J. Zhu, Y. Wang, J. Lin, D. Xia, A. Cao, Electro-and photodriven phase change composites based on wax-infiltrated carbon nanotube sponges, ACS Nano 6 (2012) 10884−10892. https://doi.org/10.1021/nn304310n

[23] H. Zhang, Q. Sun, Y. Yuan, Z. Zhang, X. Cao, A novel form stable phase change composite with excellent thermal and electrical conductivities, Chem. Eng. J. 336 (2018) 342−351. https://doi.org/10.1016/j.cej.2017.12.046

[24] A. Karaipekli, A. Biçer, A. Sarı, V.V. Tyagi, Thermal characteristics of expanded perlite/paraffin composite phase change material with enhanced thermal conductivity

using carbon nanotubes, Energy Convers. Manage. 134 (2017) 373−381. https://doi.org/10.1016/j.enconman.2016.12.053

[25] H. Ke, M.U.H. Ghulam, Y. Li, J. Wang, B. Peng, Y. Cai, Q. Wei, Ag-coated polyurethane fibers membranes absorbed with quinary fatty acid eutectics solid-liquid phase change materials for storage and retrieval of thermal energy, Renew. Energy 99 (2016) 1−9. https://doi.org/10.1016/ j.renene.2016.06.033

[26] W.T. Wang, B.T. Tang, B.Z. Ju, Z.M. Gao, J.H. Xiu, S.F. Zhang, Fe_3O_4-functionalized graphene nanosheet embedded phase change material composites: efficient magnetic-and sunlight driven energy conversion and storage, J. Mater. Chem. A, 5 (2017) 958−968. https://doi.org/10.1039/C6TA07144A

[27] K.P. Chen, X.J. Yu, C.R. Tian, J.H. Wang, Preparation and characterization of form-stable paraffin/polyurethane composites as phase change materials for thermal energy storage, Energy Convers. Manage. 77, (2014) 13−21. https://doi.org/10.1016/j.enconman.2013.09.015

[28] Y. Hong, Preparation of polyethylene-paraffin compound as a form-stable solid-liquid phase change material, Sol. Energy Mater. Sol. Cells, 64 (2000) 37−44. https://doi.org/10.1016/S0927-0248(00)00041-6

[29] M.C. Li, M.D. Jean, J.H. Chou, B.T. Lin, Effect of different amounts of surfactant on characteristics of nanoencapsulated phase change materials, Mater. Sci. Forum, 675 (2011) 541. https://doi.org/10.1007/s00289-011-0492-1

[30] F. Tang, L.K. Liu, G. Alva, Y.T. Jia, G.Y. Fang, Synthesis and properties of microencapsulated octadecane with silica shell as shape-stabilized thermal energy storage materials, Sol. Energy Mater. Sol. Cells, 160 (2017) 1−6. https://doi.org/10.1016/j.solmat.2016.10.014

[31] S. Yu, S. Wang, D. Wu, Microencapsulation of N-octadecane phase change material with calcium carbonate shell for enhancement of thermal conductivity and serving durability: synthesis, microstructure, and performance evaluation, Appl. Energy, 114, (2014) 632−643. https://doi.org/10.1016/j.apenergy.2013.10.029

[32] L. Chai, X. Wang, D. Wu, Development of bifunctional microencapsulated phase change materials with crystalline titanium dioxide shell for latent-heat storage and photocatalytic effectiveness, Appl. Energy, 138, (2015) 661−674. https://doi.org/10.1016/j.apenergy.2014.11.006

[33] T. Nomura, K. Tabuchi, C.Y. Zhu, N. Sheng, S. Wang, T. Akiyama, High thermal conductivity phase change composite with percolating carbon fiber network, Appl. Energy, 154 (2015) 678−685. https://doi.org/10.1016/j.apenergy.2015.05.042

[34] Y. Luan, M. Yang, Q. Ma, Y. Qi, H. Gao, Z. Wu, G. Wang, Introduction of an organic acid phase changing material into metalorganic frameworks and the study of its thermal properties, J. Mater. Chem. A, 4 (2016) 7641−7649. https://doi.org/10.1039/C6TA01676F

[35] T. Qian, H. Li, X. Min, W. Guan, Y. Deng, L. Ning, Enhanced thermal conductivity of peg/diatomite shape-stabilized phase change materials with ag nanoparticles for thermal energy storage. J. Mater. Chem. A, 3 (2015) 8526−8536. https://doi.org/10.1039/C5TA00309A

[36] S. Ye, Q. Zhang, D. Hu, J. Feng, Core-shell-like structured graphene aerogel encapsulating paraffin: shape-stable phase change material for thermal energy storage, J. Mater. Chem. A, 3 (2015) 4018− 4025. https://doi.org/10.1039/C4TA05448B

[37] X. Fang, P. Hao, B. Song, C.C. Tuan, C.P. Wong, Z.T. Yu, Form-stable phase change material embedded with chitosan-derived carbon aerogel, Mater. Lett. 195, (2017) 79−81. https://doi.org/10.1016/j.matlet.2017.02.075

[38] D. Yu, E. Nagelli, F. Du, L. Dai, Metal-free carbon nanomaterials become more active than metal catalysts and last longer, J. Phys. Chem. Lett. 1 (2010) 2165−2173. https://doi.org/10.1021/jz100533t

[39] A. Li, C. Dong, W. Dong, D.G. Atinafu, H.Y. Gao, X. Chen, G. Wang, Hierarchical 3D reduced graphene porous-carbon-based pcms for superior thermal energy storage performance, ACS Appl. Mater. Inter. 10 (2018) 32093−32101. https://doi.org/10.1021/acsami.8b09541

[40] S. Chandrasekaran, P.G. Campbell, T.F.Baumann, M.A. Worsley, Carbon aerogel evolution: allotrope, graphene-inspired, and 3D-printed aerogels, J. Mater. Res. 32 (2017) 4166−4185. https://doi.org/10.1557/jmr.2017.411

[41] X.C. Gui, J.Q. Wei, K.L. Wang, A.Y. Cao, H.W. Zhu, Y. Jia, Q. K. Shu, D.H Wu, Carbon nanotube sponges, Adv. Mater. 22 (2010) 617−621. https://doi.org/10.1002/adma.200902986

[42] G.Q. Zu, J. Shen, L.P. Zou, F. Wang, X.D. Wang, Y.W. Zhang, X.D. Yao, Nanocellulose-derived highly porous carbon aerogels for supercapacitors, Carbon, 99 (2016) 203−211. https://doi.org/10.1016/j.carbon.2015.11.079

[43] M.M. Titirici, R.J. White, N. Brun, V.L. Budarin, D.S. Su, F. del Monte, J.H. Clark, M.J. MacLachlan, Sustainable carbon materials, Chem. Soc. Rev. 44 (2015) 250−290. https://doi.org/10.1039/C4CS00232F

[44] A.C. Pierre, A. Rigacci, SiO_2 Aerogels, M.A. Aegerter, N. Leventis, M.M. Koebel (Eds.), Aerogels Handbook, Springer: New York, NY, USA, 2011, pp. 932

[45] T. Wu, J. Dong, F. Gan, Y. Fang, X. Zhao, Q. Zhang, Low dielectric constant and moisture-resistant polyimide aerogels containing trifluoromethyl pendent groups. Appl. Surf. Sci. 440 (2018) 95−605. https://doi.org/10.1016/j.apsusc.2018.01.132

[46] N. Leventis, C. Sotiriou-Leventis, N. Chandrasekaran, S. Mulik, Z.J. Larimore, H. Lu, G. Churu, J.T. Mang, Multifunctional polyurea aerogels from isocyanates and water, a structure−property case study, Chem. Mater. 22 (2010) 6692−6710. https://doi.org/10.1021/cm102891d

[47] S. Salimian, A. Zadhoush, M. Naeimirad, R. Kotek, S. Ramakrishna, A review on aerogel: 3D nanoporous structured fillers in polymer-based nanocomposites, Polym. Compos. 39 (2018) 3383−3408. doi.org/10.1002/pc.24412

[48] L. Zuo, Y. Zhang, L. Zhang, Y.E. Miao, W. Fan, T. Liu, Polymer/carbon-based hybrid aerogels: preparation, properties and applications, Materials, 8 (2015) 6806−6848. https://doi.org/10.3390/ma8105343

[49] N. Husing, U. Schubert, Aerogels. Ullmann's encyclopedia of industrial chemistry; John Wiley and Sons: Hoboken, NJ, USA, 2005

[50] P. Lorjai, T. Chaisuwan, S. Wongkasemjit, Porous Structure of Polybenzoxazine-Based Organic aerogel prepared by sol−gel process and their carbon aerogels, J. Sol-Gel Sci. Technol. 52 (2009) 56−64. https://doi.org/10.1007/s10971-009-1992-4

[51] L.C. Klein, Conventional Energy Sources and Alternative Energy Sources and the Role of Sol-Gel Processing, M.A. Aegerter, M. Prassas (Eds.), Sol-gel processing for conventional and alternative energy; Springer: Boston, MA, USA, 2012, pp. 1-5.

[52] C. Wingfield, L. Franzel, M.F. Bertino, N. Leventis, Fabrication of functionally graded aerogels, cellular aerogels and anisotropic ceramics, J. Mater. Chem. 21 (2011) 11737. https://doi.org/10.1039/C1JM10898K

[53] X. Pang, J. Zhu, T. Shao, X. Luo, L. Zhang, Facile fabrication of gradient density organic aerogel foams via density gradient centrifugation and uv curing in one-step. J. Sol-Gel Sci. Technol. 85 (2018) 243−250. https://doi.org/10.1007/s10971-017-4529-2

[54] Gui, J. Y.; Zhou, B.; Zhong, Y. H.; Du, A.; Shen, J. Fabrication of gradient density SiO_2 aerogel. J. Sol-Gel Sci. Technol. 2011, 58 (2), 470−475. https://doi.org/10.1007/s10971-011-2415-x

[55] F. Hemberger, S. Weis, G. Reichenauer, H.P. Ebert, Thermal transport properties of functionally graded carbon aerogels, Int. J. Thermophys. 30 (2009) 1357−1371. https://doi.org/10.1007/s10765-009-0616-0

[56] S. Koide, K. Yazawa, N. Asakawa, Y. Inoue, Fabrication of functionally graded bulk materials of organic polymer blends by uniaxial thermal gradient, J. Mater. Chem. 17 (2007) 582–590. https://doi.org/10.1039/B614001G

[57] A.C. Pierre, G.M. Pajonk, Chemistry of aerogels and their applications, Chem. Rev. 102 (2002) 4243–4265. https://doi.org/10.1021/cr0101306

[58] W. Fan, L. Zuo, Y. Zhang, Y. Chen, T. Liu, Mechanically strong polyimide/carbon nanotube composite aerogels with controllable porous structure, Compos. Sci. Technol. 156 (2018) 186–191. https://doi.org/10.1016/j.compscitech.2017.12.034

[59] R. Xu, W. Wang, D. Yu, A novel multilayer sandwich fabric- based composite material for infrared stealth and super thermal insulation protection at present, infrared stealth materials for advanced detection and stealth, Compos. Struct. 212 (2019) 58–65. https://doi.org/10.1016/j.compstruct.2019.01.032

[60] G. Jia, Z. Li, P. Liu, Q. Jing, Preparation and characterization of aerogel/expanded perlite composite as building thermal insulation material, J. Non-Cryst. Solids, 482 (2018) 192–202. https://doi.org/10.1016/j.jnoncrysol.2017.12.047

[61] Z. Qian, Z. Wang, Y. Chen, S. Tong, M. Ge, N. Zhao, J. Xu, Superelastic and ultralight polyimide aerogels as thermal insulators and particulate air filters, J. Mater. Chem. A, 6 (2018) 828–832. https://doi.org/10.1039/C7TA09054D

[62] Y. Zhang, Z. Zeng, X.Y.D. Ma, C. Zhao, J.M. Ang, B.F. Ng, M.P. Wan, S.C. Wong, Z. Wang, X. Lu, Mussel-inspired approach to cross-linked functional 3D nanofibrous aerogels for energy-efficient filtration of ultrafine airborne particles. Appl. Surf. Sci. 479 (2019) 700–708. https://doi.org/10.1016/j.apsusc.2019.02.173

[63] Y.G. Zhang, Y.J. Zhu, Z.C. Xiong, J. Wu, F. Chen, Bioinspired ultralight inorganic aerogel for highly efficient air filtration and oil–water separation, ACS Appl. Mater. Inter. 10 (2018) 13019–13027. https://doi.org/10.1021/acsami.8b02081

[64] Y.B. Pottathara, V. Bobnar, M. Finsgar, Y. Grohens, S. Thomas, V. Kokol, Cellulose nanofibrils-reduced graphene oxide xerogels and cryogels for dielectric and electrochemical storage applications. Polymer 147 (2018) 60–270. https://doi.org/10.1016/j.polymer.2018.06.005

[65] R. Sun, H. Chen, Q. Li, Q. Song, X. Zhang, Spontaneous assembly of strong and conductive graphene/polypyrrole hybrid aerogels for energy storage, Nanoscale 6 (2014) 12912–12920. https://doi.org/10.1039/C4NR03322A

[66] M. Wang, I.V. Anoshkin, A.G. Nasibulin, J.T. Korhonen, J. Seitsonen, J. Pere, E.I. Kauppinen, R.H.A. Ras, O. Ikkala, Modifying native nanocellulose aerogels with

carbon nanotubes for mechanoresponsive conductivity and pressure sensing, Adv. Mater. 25 (2013) 2428–2432. https://doi.org/10.1002/adma.201300256

[67] R. Begag, S. White, J.E. Fesmire, W.L. Johnson, Hybrid aerogel-MLI insulation system performance studies for cryogenic storage in space applications. MRS Proc. 1306 (2011) mrsf10-1306- bb01-03. https://doi.org/10.1557/opl.2011.88

[68] C. Shan, L. Wang, Z. Li, X. Zhong, Y. Hou, L. Zhang, F. Shi, Graphene oxide enhanced polyacrylamide-alginate aerogels catalysts, Carbohydr. Polym. 203 (2019) 19–25. https://doi.org/10.1016/j.carbpol.2018.09.024

[69] S. Zhao, W.J. Malfait, N. Guerrero-Alburquerque, M.M. Koebel, Nystrom, G. Biopolymer Aerogels and foams: chemistry, properties, and applications, Angew. Chem. Int. Ed. 57 (2018) 7580–7608. https://doi.org/10.1002/anie.201709014

[70] Z. Liu, M.A. Meyers, Z. Zhang, R.O. Ritchie, Functional gradients and heterogeneities in biological materials: design principles, functions, and bioinspired applications, Prog. Mater. Sci. 88 (2017) 467–498. https://doi.org/10.1016/j.pmatsci.2017.04.013

[71] T.D. Nguyen, D. Tang, F.D' Acierno, C.A. Michal, M.J. Maclachlan, Biotemplated lightweight γ-alumina aerogels, Chem. Mater. 30 (2018) 1602–1609. https://doi.org/10.1021/acs.chemmater.7b04800

[72] A. Lamy-Mendes, R.F. Silva, L. Duraes, Advances in carbon nanostructure-silica aerogel composites: a review, J. Mater. Chem. A, 6 (2018) 1340–1369. https://doi.org/10.1039/C7TA08959G

[73] R. Baetens, B.P. Jelle, A. Gustavsen, Aerogel insulation for building applications: a state-of-the-art review, Energy Build. 43 (2011) 761–769. https://doi.org/10.1016/j.enbuild.2010.12.012

[74] A. Shinko, S.C. Jana, M.A. Meador, Crosslinked polyurea-co- polyurethane aerogels with hierarchical structures and low stiffness, J. Non-Cryst. Solids, 487 (2018) 19–27. https://doi.org/10.1016/j.jnoncrysol.2018.02.020

[75] M. Aghabararpour, M. Mohsenpour, M. Motahari, A. Abolghasemi, Mechanical properties of isocyanate crosslinked resorcinol formaldehyde aerogels, J. Non-Cryst. Solids 481 (2018) 548–555. https://doi.org/10.1016/j.jnoncrysol.2018.02.020

[76] I.A. Principe, A.J. Fletcher, Parametric study of factors affecting melamine-resorcinol- formaldehyde xerogels properties, Mater. Today Chem. 7 (2018) 5–14. https://doi.org/10.1016/j.mtchem.2017.11.002

[77] H. Chen, X. Li, M. Chen, Y. He, H.S.C. Zhao, Self-crosslinked melamine-formaldehyde-pectin aerogel with excellent water resistance and flame retardancy,

Carbohydr. Polym. 206 (2019) 609– 615.
https://doi.org/10.1016/j.carbpol.2018.11.041

[78] I.A. Principe, B. Murdoch, J.M. Flannigan, A.J. Fletcher, Decoupling microporosity and nitrogen content to optimize CO_2 adsorption in melamine resorcinol formaldehyde xerogels. Mater. Today Chem. 10 (2018) 195−205.
https://doi.org/10.1016/j.mtchem.2018.09.006

[79] M.A.B.Meador, C.R. Aleman, K. Hanson, N. Ramirez, S.L. Vivod, N. Wilmoth, L. McCorkle, Polyimide aerogels with amide cross-links: a low-cost alternative for mechanically strong polymer aerogels, ACS Appl. Mater. Inter. 7 (2015) 1240−1249.
https://doi.org/.org/10.1021/am507268c

[80] Y. Sun, L. Xia, J. Wu, S. Zhang, X. Liu, Mesoscale self- assembly of reactive monomicelles: general strategy toward phloroglucinol-formaldehyde aerogels with ordered mesoporous structures and enhanced mechanical properties, J. Colloid Interface Sci. 532 (2018) 77−82. https://doi.org/10.1016/j.jcis.2018.07.104

[81] Z. Niu, W. Yuan, Highly efficient thermo-and sunlight-driven energy storage for thermo-electric energy harvesting using sustainable nanocellulose-derived carbon aerogels embedded phase change materials, ACS Sustain. Chem. Eng. 7 (2019) 17523−17534. https://doi.org/10.1021/acssuschemeng.9b05015

[82] L. Zhang, L. An, Y.Wang, A. Lee, Y. Schuman, A. Ural, A.S. Fleischer, G. Feng, Thermal enhancement and shape stabilization of a phase-change energy-storage material via copper nanowire aerogel, Chem. Eng. J. 373 (2019) 857−869.
https://doi.org/10.1016/j.cej.2019.05.104

[83] Z. Fan, D.Z.Y Tng, C.X.T. Lim, P. Liu, S.T. Nguyen, P. Xiao, A. Marconnet, C.Y. Lim, H.M. Duong, thermal and electrical properties of graphene/carbon nanotube aerogels, Colloids Surf. A Physicochem. Eng. Asp. 445 (2014) 48−53.
https://doi.org/10.1016/j.colsurfa.2013.12.083

[84] B. Muand, M. Li, Synthesis of novel form-stable composite phase change materials with modified graphene aerogel for solar energy conversion and storage, Sol. Energy Mater. Sol. Cells, 191(2019) 466−475. https://doi.org/10.1016/j.solmat.2018.11.025

[85] H. Hong, Y. Pan, H. Sun, Z. Zhu, C. Ma, B. Wang, W. Liang, B. Yang, A. Li, Superwetting polypropylene aerogel supported form-stable phase change materials with extremely high organics loading and enhanced thermal conductivity, Sol. Energy Mater. Sol. Cells, 174(2018) 307−313. https://doi.org/10.1016/j.solmat.2017.09.026

[86] F. Li, L. Xie, G. Sun, Q. Kong, F. Su, Y. Cao, J. Wei, A. Ahmad, X. Guo, C.M. Chen, Resorcinol-formaldehyde based carbon aerogel: preparation, structure and

applications in energy storage devices, Micropor. Mesopor. Mater. 279 (2019)
293–315. https://doi.org/10.1016/j.micromeso .2018.12.007

[87] J. Shen, P. Zhang, L. Song, J. Li, B. Ji, J. Li, L. Chen, Polyethylene glycol supported
by phosphorylated polyvinyl alcohol/graphene aerogel as a high thermal stability phase
change material, Compos. Part B: Eng. 179 (2019) 107545.
https://doi.org/10.1016/j.compositesb. 2019.107545

[88] L.S. Tang, J. Yang, R.Y. Bao, Z.Y. Liu, B.H. Xie, M.B. Yang, W. Yang,
Polyethylene glycol/graphene oxide aerogel shape-stabilized phase change materials
for photo-to-thermal energy conversion and storage via tuning the oxidation degree of
graphene oxide, Energy Con. Manag. 146 (2017) 253–264.
https://doi.org/10.1016/j.enconman.2017.05.037

[89] X. Huang, W. Xia, R. Zou, Nanoconfinement of phase change materials within
carbon aerogels: phase transition behaviours and photo-to-thermal energy storage, J.
Mater. Chem. A, 2(2014) 19963–19968. https://doi.org/10.1039/C4TA04605F

[90] S.T. Nguyen, H.T. Nguyen, A. Rinaldi, N. P. Nguyen, Z. Fan, H.M. Duong,
Morphology control and thermal stability of binderless-graphene aerogels from
graphite for energy storage applications, Colloids Surf. A Physicochem. Eng. Asp. 414
(2012) 352–358. https://doi.org/10.1016/j.colsurfa.2012.08.048

[91] J. Wang, D. Liu, Q. Li, C. Chen, Z. Chen, P. Song, J. Hao, Y. Li, S. Fakhrhoseini,
M. Naebe, X. Wang, Lightweight, superelastic yet thermoconductive boron nitride
nanocomposite aerogel for thermal energy regulation, ACS Nano, 13 (2019)
860–7870. https://doi.org/10.1021/acsnano.9b02182

[92] J. Yang, G.Q. Qi, R.Y. Bao, K. Yi, M. Li, L. Peng, Z. Cai, M.B. Yang, D. Wei, W.,
Yang, Hybridizing graphene aerogel into three-dimensional graphene foam for high-
performance composite phase change materials, Energy Stor. Mater. 13(2018) 88–95.
https://doi.org/10.1016/j.ensm.2017.12.028

[93] J. Yang, G.Q. Qi, Y. Liu, R.Y. Bao, Z.Y. Liu, W. Yang, B.H. Xie, M.B. Yang,
Hybrid graphene aerogels/phase change material composites: thermal conductivity,
shape-stabilization and light-to-thermal energy storage, Carbon, 100 (2016) 693–702.
https://doi.org/10.1016/j.carbon.2016.01.063

[94] Z. Niu, W. Yuan, Highly efficient thermo-and sunlight-driven energy storage for
thermo-electric energy harvesting using sustainable nanocellulose-derived carbon
aerogels embedded phase change materials, ACS Sustain. Chem. Eng. 7 (2019)
17523–17534. https://doi.org/10.1021/acssuschemeng.9b05015

[95] H. Im, T. Kim, H. Song, J. Choi, J.S. Park, R. Ovalle-Robles, H.D. Yang, K.D.
Kihm, R.H. Baughman, H.H. Lee, T.J. Kang, High-efficiency electrochemical thermal

energy harvester using carbon nanotube aerogel sheet electrodes, Nat. Commun. 7(2016) 10600. https://doi.org/10.1038/ncomms10600

[96] J. Zhao, W. Luo, J.K. Kim, J. Yang, Graphene-oxide aerogel beads filled with phase change material for latent heat storage and release, ACS Appl. Energy Mater. 2 (2019) 3657–3664. https://doi.org/10.1021/acsaem.9b00374

[97] L. Liu, K. Zheng, Y. Yan, Z. Cai, S. Lin, X. Hu, Graphene aerogels enhanced phase change materials prepared by one-pot method with high thermal conductivity and large latent energy storage, Sol. Energy Mater. Sol. Cells, 185 (2018) 487–493. https://doi.org/10.1016/j.solmat. 2018.06.005

[98] X. Wang, G. Li, G. Hong, Q. Guo, X. Zhang, Graphene aerogel templated fabrication of phase change microspheres as thermal buffers in microelectronic devices, ACS Appl. Mater. Inter., 9 (2017) 41323–41331. https://doi.org/10.1021/acsami.7b13969

[99] K. Liang, L., Shi, J. Zhang, J. Cheng, X. Wang, Fabrication of shape-stable composite phase change materials based on lauric acid and graphene/graphene oxide complex aerogels for enhancement of thermal energy storage and electrical conduction, Thermochim. Acta, 664 (2018) 1–15. https://doi.org/10.1016/j.tca.2018.04.002

[100] H. Zhou, D. Zhang, Effect of graphene oxide aerogel on dehydration temperature of graphene oxide aerogel stabilized $MgCl_2. 6H_2O$ composites, Sol. Energy, 184 (2019) 202–208. https://doi.org/10.1016/j.solener.2019.03.076

Aerogels II: Preparation, Properties and Applications   Materials Research Forum LLC
Materials Research Foundations **97** (2021) 145-167   https://doi.org/10.21741/9781644901298-8

Chapter 8

Aerogels for Sensor Application

Sapna Raghav[1], Pallavi Jain[2], Praveen Kumar Yadav[3], Dinesh Kumar[4,*]

[1]Department of Chemistry, Banasthali Vidyapith, Banasthali, Tonk 304022, Rajasthan, India

[2]Department of Chemistry, SRM Institute of Science & Technology, Delhi-NCR Campus, Modinagar-210204, India

[3]Academy of Scientific and Innovative Research, CSIR-National Physical Laboratory, Dr K.S. Krishnan Marg, New Delhi-110012, India

[4]School of Chemical Sciences, Central University of Gujarat, Gandhinagar, India

*dinesh.kumar@cug.ac.in

Abstract

Aerogels with air-filled pores and interconnected 3D solid networks show unique characteristics and, therefore, have tremendous applications in various fields. Integrating specific characteristics of aerogels, large surface area, low density, and high porosity are included which opens up possibilities for new application areas. Aerogels' advanced features provide high selectivity and sensitivity, fast recovery and response to sensing materials in sensors such as biosensors, gas, pressure, and strain sensors. In recent years significant work has been done regarding the development of aerogel-based sensors. In this chapter, recent challenges and some approaches to high-performance aerogel-based sensor development are summarized.

Keywords

Aerogel, Sensors, Stress, Electrochemical Sensors, Strain

Contents

1. Introduction

Aerogels are considered as light solid materials. They comprise macropores and mesopores and have a porosity percentage which is greater than 50% in a nanoscale network. These pores are connected to each other through the particles that are bonded covalently and then loaded with gas. The aerogels are prevalent and have attributes like high porosity and a more specific area alongside low thermal conductivity and less density [1, 2].

S. Kistler established the term aerogel in 1932 [3]. It was useful as a gel in which gas substituted the liquid part without breaking the solid framework of gel. Hence, an aerogel can be called a solid gel that has micropores and comprises gas in a dispersed phase. Aerogels are porous and light because its volume is made 95% of air, and the rest of the part is solid. These are nanoporous materials of an exceptional kind and comprise distinct attributed, like lesser bulk density (0.003-0.5 gcm^{-1}), large porosity, and huge surface area

(500-1500 m^2g^{-1}). The refractive index of air is 1.004 $Wm^{-1}k^{-1}$, while that of aerogels is 1.007-1.24 $Wm^{-1}k^{-1}$. Similarly, the thermal conductivity of air is 0.0209 $Wm^{-1}k^{-1}$, and in comparison, the thermal conductivity of aerogel is 0.02 $Wm^{-1}k^{-1}$ [4].

2. Classification and physicochemical properties of aerogels

Recently, various modified processes have been developed to prepare different aerogels. The aerogels that are formed are categorized based on their structure, composition, attributes, and implementations in various fields. The aerogels are categorized as inorganic, organic, and hybrid aerogels.

2.1 Inorganic aerogels

Inorganic aerogels have 3D porous networks and are composed of inorganic materials, like carbon-based aerogels, oxide-based aerogels, chalcogenide aerogels, and metallic aerogels. These aerogels are easily available and their properties are based on the inorganic material.

2.2 Oxide-based aerogels

The 3D porous networks of aerogels are made of oxide and are synthesized via the process of hydrolysis of equivalent precursors. The world's first aerogel was recorded by Kistler (oxide-based aerogel) and was later improvised by Nicolaon and Teichner on the preparation process [5, 6]. Since then, both the preparations processes were reported. More and more attention has been paid to oxide-based aerogels in which silicon dioxide (SiO_2), titanium dioxide (TiO_2), zirconium dioxide (ZrO_2), vanadium pentoxide (V_2O_5) and aluminium oxide (Al_2O_3) are studied intensively and commonly used. Apart from the low density, less thermal conductivity along with more specific surface area (SSA), oxide-based aerogels also have good thermal stability and non-flammability. Such aerogels can therefore, be useful in thermal isolation at high temperatures and show excellent use in buildings. In 2003, Mars exploration rovers used SiO_2 aerogel having a much lesser thermal conductivity and great non-flammability for thermal insulation. The gradient aerogel combining the SiO_2 aerogel having a structure of gradient density was used in the 2006 stardust mission to capture the particles of high velocity in outer space, because of their distinct nanostructure; Al_2O_3 aerogel show increased high temperature structural stability in contrast to SiO_2 aerogel. The shrinkage in SiO_2 happens at a temperature above 600 °C because of sintering. Al_2O_3 aerogel is also a prevalent support for high-temperature catalysts. Like Al_2O_3, ZrO_2 aerogels also have chemical and structural stability at high temperatures and are also a support for catalysts at high temperatures. ZrO_2 aerogel revealed excellent capability in solid oxide fuel cells. TiO_2 aerogel, in particular, anatase TiO_2 aerogel, has photovoltaic and photocatalytic properties and has been used in many fields. V_2O_5 is a layered compound and can increase lithium-ion battery capacity as it can host two Li-ions. Hence, these are being used widely as cathode material. V_2O_5 is material for catalysis, and the aerogel derived from it can be used efficiently as a compound for selective catalysis [7-14].

2.3 Metallic aerogels (MAgs)

Initially, the metal-oxide aerogels were subjected to carbothermal reduction, resulting in the formation of metal/carbon (M/C) composite aerogels instead of pure MAgs. M/C aerogels high-temperature post-treatment was built-up to eliminate the carbon for preparing pure MAg, but the actual surface area of the aerogel obtained is significantly reduced. In the meantime, an easy approach was proposed to prepare aerogel-like nanoparticles (NPs) having a hundred nanometers diameter, and to aggregate the NPs, dithiol was utilized as a linker to obtain a 3D network of NPs. Afterwards, fine nanostructure metal aerogels were derived from the metal NPs colloidal solution through the gelation method (two-step process), derived. The synthesis has been made facile and has become a one-step process. MAgs can also be produced from the direct freezing technique implemented on the suspension of metal NPs and nanowires. MAgs have shown fine and a better structure with pores along with extraordinary catalytic performance and a better specific area. These aerogels also have brilliant electrical conductivity demonstrating a good prospect in electronics. MAgs that comprise distinct metals have their own exceptional attributes. Platinum-based aerogels are multifaceted electrochemical catalysts showing amazing catalytic activity in the oxygen (O_2) reduction, formic acid oxidation, and many more. The palladium aerogels show brilliant performance in catalysis of alcohol and formic acid oxidation. Gold based aerogels (Au-Ags) also show the prospect of glucose sensors. Au-Ags show convenient optical transparency, surface area, and porosity by changing the quantity of oxidant used in the process of gelation. Also, the freeze-drying in directions, the resultant Au-Ags, demonstrated a distinct anisotropic microporous structure. It can be thus used widely in electronic devices. Copper-based aerogels can also be useful in catalyzing carbon dioxide (CO_2) reduction and bio-sensing [15-25].

2.4 Carbon aerogels (CAgs)

The carbon aerogels (CAgs) were firstly manufactured by using a pyrolysis process in which resorcinol-formaldehyde (RF), organic aerogels were thermally treated at high temperature. Diverse CAgs utilizing distinct organic and biomass pioneers were successfully prepared by using this method. The utilization of graphene and carbon nanotubes (CNTs) has brought a new direction to the synthesis of CAgs because of their amazing properties. The CNTs and graphene aerogels (GAgs) can be bestowed good flexibility, stretchability, and compressibility using structural design. The better chemical and thermal stability of graphene and CNTs aerogels is because of a high degree graphitization. They also have better electrical properties in contrast to regular CAgs and is useful in many sectors. The electrically conductive and porous 3D network intended for effective mass transportation and electron transfer can be provided by CAgs [26-32].

2.5   Chalcogenide aerogels

The synthesis of such aerogels is induced by the evolution of semiconductor NPs. The dielectric matrixes having nanostructure, and large and quantum sized semiconductors, was designed by utilizing cadmium sulfide (CdS, first pioneer chalcogenide aerogel) in 1997. The mechanism of aggregation was studied later to form transparent gels as well as dense precipitates in CdS colloidal solutions. Later on, CdS gel was reproduced with the help of supercritical drying (SCD). A general strategy was reported that involved aggregation using NP oxidation in the colloidal solution to create chalcogenide aerogels series like CdS, zinc sulfide (ZnS), cadmium selenium (CdSe), lead sulfide (PbS), cadmium tellurium (CdTe), and lead selenium (PbSe). The materials used for the production of these types of aerogels possess excellent semiconductor properties. Thus, the obtained aerogels were taken the intrinsic attributes and provide an effective implementation in photovoltaics, photocatalysis, electrocatalysis, sensors, and LEDs [33-35].

2.6   Organic aerogels (OAgs)

The organic aerogels (OAgs) like RF aerogel have a 3D porous network comprising macromolecules. It is used as a precursor for carbon. For the manufacturing of RF aerogels, the conditions for preparing such as reaction temperature, monomers, solvent, catalysts, and solvent were made optimum. In addition, the gradient RF aerogels can be obtained by layer-by-layer gelation and layered sol co-gelation process. The important and essential OAg is cellulose aerogel, which is decomposable, biologically compatible, sustainable, and renewable. Cellulose aerogels are generally prepared by dissolving and regenerating cellulose in a definite solvent, accompanied by freezing or SCD depending on the solvent used. The aqueous solution of lithium hydroxide (LiOH) or sodium hydroxide (NaOH)/urea, lithium chloride (LiCl)/N, N-dimethylacetamide, ionic liquids, cuprammonium solution, are some effective cellulose solvent systems. Among these, the aqueous solution of NaOH/urea is cheap, eco-friendly, and suitable. The resulting cellulose aerogels are biologically compatible, have better mechanical strength, and is also useful as an element of support. There are numerous OAgs which have been reported earlier like cyclodextrin, chitosan, polyurethane (PU), polyvinylpolymethyl siloxane (PVPMS), polyvinylalcohol (PVA), polypyrrole (PPy), and kevlar. It is easy to control the properties of OAgs by changing the monomer [36-42].

Materials Research Forum LLC

https://doi.org/10.21741/9781644901298-8

2.7   Hybrid aerogels (HAgs)

Hybrid aerogels (Hags) obtained by mixing compounds together have superior properties owing to their synergistic effect. The objective of improvising the mechanical characteristics of silica aerogels forces the growth of HAgs. It is possible to prepare the Hags consisting of the compounds as long as each compound can gelatinize itself. The Hags have therefore expanded speedily, covering almost all types of aerogels.

Hags preparation has now proved to be an effective and potential method of overcoming the cons of single-compound aerogels. The Hags show brilliant sensing performance [43-45].

3.   Sensors application

The sensors are those devices that can convert external stimuli into data for communication after reacting with them. Advanced sensing materials are the main elements, and they need to be used efficiently. According to the usage, the sensors are categorized as biosensors, pressure sensors, gas sensors, and also other types.

3.1   Gas sensors

Gas sensors have become increasingly vital in a lot of fields like detecting poisonous and exothermic gases to ensure public safety and monitoring industry. Resistive gas sensors are broadly employed as these are economical, facile approach of processing, and simple operation. The sensor's conductivity will alter as the gas molecules are taken up by the active sensing layer surface. There is a strong concern about the stability, selectivity, sensitivity, recovery rate, detection limit, and the response of these types of sensors. The reasons which support the use of aerogel-based materials for gas sensing are a high ration of a surface-to-volume ratio and a SSA that provides sufficient surfaces for gas adsorption and a good number of active sites for gas sensing. The 3D porous structure of the material also provides rapid and steady transportation for the gas diffusion. As a result, the aerogels possess low detection limits, a rapid rate of recovery and response, along with better sensitivity. Si-doped on an Au-patterned Al_2O_3 substrate shows good stability and can sense humidity and also exhibit a better ability to reproduce. The sensitivity is also high in contrast to the xerogel produced in ambient drying. The thickness of the film affects the working of the final sensor. The thinner film possessed a better rate of response, whereas the thicker film showed a better rate of recovery. Various oxide-based aerogels were subsequently produced for gas detection. The better sensing efficiency (for gases) of semiconducting oxides is demonstrated by the resulting aerogels having smaller grain size and high porosity when combined with the 3D network [46-51]. Some of aerogel sensors are tabulated in Table 1 [50].

Table 1 Aerogel-based sensors [50] [With permission taken from WILEY-VCH Verlag GmbH & Co. KGaA, Weinheim].

Aerogel sensing material	Analyte	Sensing range	Response or recovery time and detection limit
Si film	Humidity	20-90% RH	41/55 s
TiO_2	Humidity	40-80% RH	<2/14 s
Graphene	Ammonia (NH_3), NO_2	0.02-85 ppm, 0.1-1 ppm	100/500 s, 10 ppb 116/169 s, 50 ppb
CQDs/Si	NO_2	2-10 ppm	250 ppb
CuO/SnO_2	NO_x	1-100 ppm	7-26 s, 1 ppm
SnO_2/Graphene	NO_2	10-200 ppm	190/224 s, 10 ppm
Fe_3O_4/Graphene	NO_2	30-400 ppm	275/738 s, 30 ppm
ZnO/Graphene	NO_2	10-200 ppm	132/164 s, 10 ppm
MoS_2	-	50-500 ppb	<40/<40 s, 50 ppb
MoS_2/Graphene	-	50-1000 ppb	<1/1 min, 50 ppb
WS_2/Graphene	-	0.05-2 ppm	100/300 s, 10-15 ppb
TiO_2/SiO_2	H_2S	0.5-50 ppm	53/74 s, 0.5 ppm
Pt/Graphene	H_2	2000-20000 ppm	0.97/0.72 s, 500 ppm
MCNT	Chloroform	-	0.5/0.5 s, 1 ppb

ZnO aerogel (hierarchical) that is made of hollow nanospheres (NSs) in which the hollow walls are considered as the depletion layer is highly vulnerable to target gas. It endows this hierarchical aerogel has extraordinary sensing efficiency at low-temperature in contrast to the solid NSs aerogels. Hollow TiO_2 nanotube aerogels were prepared by deposition of that has a basis on aerogel based on nanocellulose. By atomic layer deposition (ALD), well-dispersed TiO_2 layers on nanocellulose aerogel nanofibrils are easily prepared, and the hollow nanotubes of TiO_2 aerogels can be obtained by accompanying calcinations (450

°C, 8 hours) to decompose nanocellulose template. These hollow TiO_2 nanotube aerogels possess excellent humidity sensing efficiency because of their hierarchical structure. In addition, TiO_2 can be deposited on silica aerogel via ALD and the composite aerogel that has a mismatch in lattice among them. The TiO_2 grain size received can provide more active sites for sensing of hydrogen sulfide (H_2S). Aluminium doped zinc oxide aerogels can be prepared by gas-phase synthesis. The detection limit obtained was low. CuO-doped aerogel demonstrated good restorability along with brilliant selectivity at a lower concentration of gas to detect oxides of nitrogen (NO_x) [52-54].

For the first time, functionalized silica aerogels having strong carbon quantum dots (CQDs) (fluorescent) are prepared and used for easy, responsive, and selective sensing of nitrogen oxide (NO_2) gas. Homemade silica aerogels having 801.17 m^2g^{-1} SSA were functionalized in ethanol with branched polyethylenimine-capped CQDs with a quantum yield of fluorescence greater than 40 %. The obtained porous hybrid material of CQD-aerogel could maintain in its solid-state its excellent fluorescence activity, which is quenched sensitively by NO_2 gas and suggested use of the novel fluorescent-functional aerogels in gas detection [55].

3.2   Gas-phase sensing

Aerogels based on silica have many characteristics that make them most suitable as gas sensors. The porosity percentage is high around 99%. Hence these aerogels are mostly gases in nature based on their volume. The refractive index is almost similar to air and hence guarantees the quick flow of light across the volume of the aerogel. The internal surface area is large for the analytic sorption and also to enable the linking of the functional groups so that they bring the analyses closer to the aerogel and to detect them. Based on the analyte nature and the surrounding within which they are being used, the surfaces are framed as hydrophobic or hydrophilic. Owing to their synthesis through gelation process, the approach to add various chromophores, biological receptors, and fluorophores inside the structure is facile. This restricts the range of potential in implementations only by the thought process of engineers and scientists. They can be manufactured as powder, monolithic form, a thin layer, and this permits a good flexible nature. Also, they are responsive to the process of hybridization in combination with organic polymers to provide mechanical robustness, durability, and different electrical conductivities as wanted. The most advantageous implementation of silica aerogel is considered as gas-phase sensing. The foremost approaches of the discovery of gas-phase that are applied involving silica aerogels require processes like photoluminescence, fluorescence, absorbance, circular dichroism, and Raman scattering. Methods that involve electricity, which is dependent on capacitance measurements and resistance, are also shown. Many researchers have involved the abilities of silica aerogels as gas sensors in their research. These reviews explore in depth the evolution of silica aerogel sensors to detect O_2 and humidity. The review of Wagner et al. [56] discussed different mechanisms opted to detect gas in different mesoporous materials that have a consistent and numbered pore. The research by Barczak et al. [57] summarized the gelation process and optical sensing for O_2, pH, NH_3, and CO_2 with the help of sol-gel based sensors. In this chapter, we deal with different categories of gas-phase analytes like water vapor, O_2, acids, bases, and hydrocarbons (HCs).

3.3 Water vapor sensor

A thin film of silica aerogel was prepared by Wang and co-workers [58] for sensing humidity that was dependent on signals of electrical impedance. Their method entailed the making of comb-patterned, a pair of meshed Au-film electrodes put on the substrate of alumina and deposition of silica sol over it with the help of the process of dipping. The coating was improvised by waiting for the sol to arrive in a particular range of viscosity before dipping the substrate throughout the whole gelation process. The rate of dipping was also made specific. For the completion of the aerogel process, under a highly critical environment, the sensor that was gel-coated was dried slowly. The calcination took place at 500 °C and then left foraging for 12 hours in humid conditions. This was done for the surface of rehydroxylation. Sensors were also made by letting the gel-coating to dry off using certain drying methods under immediate conditions so as to prepare xerogel films. The test for sensors also entailed measurements of impedance as a relative humidity function present between two electrodes, temperature, and frequency. The sensors coated with xerogel exhibited little impedance reaction to alterations in relative humidity, whereas the sensors coated with aerogel demonstrated a drop in impedance of 700 k ohms at a frequency of 1 kHz and at the temperature of 25 °C. The relative humidity range was from 20 to 90%. A non-linear impedance response to frequency was translated by the researchers as proof in the system for capacitance owing to the effects of space charge polarization. A conclusion was made that the thickness of aerogel and its pore structure were two main factors that affected the response of the sensor to water, and the sensor could be shaped as a resistor capacitor circuit. The protons would supply the required electrical conductivity [59, 60].

3.4   Oxygen sensor

Luminescent probes combined with monolithic silica aerogel have been useful in detecting O_2 that plays the role of suppressing luminescence. An experiment was performed wherein monolithic silica aerogels were heated using the energy of microwave within an atmosphere that was reduced. This was done to make vacancies in the structure of aerogel. There was luminance owing to the ultra-violet (UV) light in the absence of O_2. The luminescence was put off as the O_2 molecules came in touch with the silica aerogel, thus making the primary base for the detector of O_2. The aerogels were treated using HMDS so as to make the surface to repel water. This caused the sensitivity towards O_2 to decrease seven times in contrast to the aerogel that was hydrophilic. The sensitivity was the highest at partial pressures of O_2 of 100-300 torr. Monolithic silica aerogels were added with modified functions, properties, and other attributes using N-(3-trimethylsilylpropyl)-2,7-diazapyrenium bromide (DAP) using co-gelation and post-gelation techniques. The process of co-gelation required combining DAP and trimethoxysilane (TMOS) precursors, gel washing using acetone, followed by CO_2-SCD or evaporative air drying in order to develop an aerogel or a xerogel, respectively.

Owing to the formation of two simultaneous gelation products, DAP was dispersed completely. The gelation functionalization method that was followed after entailed its reception with DAP after gel aging. Again, the drying process of co-gelled resultants followed under supercritical conditions. This caused the DAP to get retained mainly on the aerogels' outer surface. Hence, the researchers came to the conclusion that the consequent gelation process would make for the quicker reaction rates and times for the detectors based on sol-gel. Fast luminescence was extinguished using O_2 during the test of functionalized-DAP gels for both aerogels. Unrealistic, slower reactive times were received for aerogel. The emission lifetimes of the aerogels modified by DAP in contrast to DAP dissolute in aq. methyl alcohol exhibited higher lifetimes for DAP modified aerogel. There was no difference among the doping in post-gelation product and co-gelled methods.

Ru(II)-phenanthroline was paired with one of the paired electron acceptors. This luminescent compound played the role of a light unit for harvesting and was used for the post gelation modification of silica aerogel. It was used as a compound for doping. Higher response rates were achieved towards O_2 in contrast to that of aerogels modified by DAP. An active sensitivity range, which was better than that of the Ru(II)-phenanthroline, was obtained.

Ru(II)-bipyridyl and Ru(II)-4,7-diphenylphenanthroline complexes were introduced into hydrophilic monolithic Si aerogels produced by the single-pot technique and contrasted the comparable responsive times of the three aerogels to O_2. The Ru(II)-bipyridyl aerogel, in contrast, was much less responsive, producing only a 10% alteration in intensity over the same range in levels of O_2 [61-63].

3.5   Pressure sensor

Aerogel is a high porosity and low-density material. However, there is quite limited research on metal-based aerogel with good conductivity, which hinders its use in electronic devices. With the fast growth of smart electronic devices, wearable pressure sensors are in great demand. Nevertheless, achieving high-performance pressure sensors with high sensitivity, a wide range of responses, and a low limit of detection simultaneously remains a huge challenge. To deal with this issue, copper nanowire-based aerogels were designed via the one-pot process, and the assembly of copper nanowire into hydrogel was investigated. For the first time, the "bubble-controlled assembly" mechanism was proposed and reported the tunable pore structure along with 4.3-7.5 mg/cm^3 densities of the nanowire aerogel. Later, flexible pressure sensors having 0.02-0.7 kPa^{-1} sensitivities were developed by using copper nanowire-based aerogels. The fabricated pressure sensors were extremely light and greatly expanded their potential for application [64]. Polyimide/CNTs composite aerogels synthesized via freeze-drying and thermal imidization, possessed high porosity, resistance against high temperature, robust, super-elastic property. These composite aerogels also displayed excellent sensing efficiency, exceptionally low limit of detection, extremely high sensitivity, extraordinary stability, good recovery, and response time along with outstanding detection efficiency toward compression, bending, and distortion. These aerogels efficiently worked in numerous harsh atmospheres. Polyimide/CNTs composite aerogels used to detect an efficiently full range of human motion and also precisely detected distribution of pressure when assembled in E-skin. These composite aerogels showed huge potential as a wearable pressure sensor with high performance [65].

Silicon nanocrystals having excellent optoelectronic properties, when incorporated in lightweight aerogels having a high porosity, resulted in a hybrid composite material having the properties of both silicon nanocrystals and aerogels. Different sized silicon nanocrystals were formulated with triethoxyvinylsilane and finally used to develop highly porous photoluminescent silica hybrids, which a potent sensor for high-energy materials [66].

High compressibility elastic carbon materials are ideal for flexible piezoresistive sensors. The excellent mechanical strength and conductivity of carbon nanotubes, making them the hopeful nano blocks to fabricate flexible and compressible carbon aerogels. The multiwalled carbon nanotubes utilizing chitosan (binder) resulted in elastic and compressible carbon aerogel, which showed stability even at 90% of compression strain and exhibited outstanding fatigue resistance. Because of excellent resistance stability and better sensitivity, these aerogels can easily sense human bio-signals. Adequate mechanical and sensing efficiency makes it possible to use piezoresistive sensors [67].

Because of the significant and extensive relevance in industries, pharmaceuticals, good stores, agriculture, humidity sensors have received a substantial interest in recent years. The humidity sensors generally had low sensitivity with complex sensing mechanism and needed labor-intensive and time taking process. Exploring an ideal sensing material to amplify humidity sensor sensitivity remained a major challenge. In this context, a highly sensitive humidity sensor, poly-N-isopropylacrylamide/AuNP hybrid aerogel, was developed via a freeze-drying process. The polymer/ AuNP aerogel possessed high sensitivity for water molecules and used to sense various states of human breathe (normal, fast, and deep) in distinct individuals (people with illness, people who smoke and people who are normal). These humidity sensors also detect whistle tune [68].

With relevance in many wearable devices along with piezo resistive sensors, 3D CAgs having a significantly high elasticity and compressibility hold great promise. MXenes come up as 3D conductive materials to develop piezoresistive CAgs. Nonetheless, owing to the low aspect ratio, continuous MXene microstructures are difficult to attain. Cellulose nanocrystals are employed for connecting nanosheets of MXene and lamellar CAgs having super-mechanical performance as well as ultra-high linear sensitivity to resolve this issue. Cellulose nanocrystals and MXene interaction resulted in an extremely stable lamellar structure consisting of the wave shape, which can suffer very high compression strain (95%) with long-term i.e., 10000 cycles at 50% strain. It even displayed a broad linear range of 50 Pa-10 kPa as well as enormously high sensitivity and also effective in sensing a minute change in pressure. Such benefits allow these CAgs to be utilized in piezo resistive wearable devices for bio-signal detection [69].

Recently, polymer aerogels are developed as potential materials that have high elasticity and sensing efficiency to temperature and pressure. The challenge, though, is to read temperature and pressure signals separately. To resolve this, a sensor having the dual parameter to decouple temperature and pressure reading was developed. These sensors are basically polymer aerogels with thermoelectric properties. The polymer aerogels composed of poly(3,4-ethylenedioxythiophene) poly(styrenesulfonate) in a polar solvent i.e., dimethylsulfoxide (DMSO) having a high boiling point reported as a dual-parameter sensing device that efficiently decouples temperature and pressure without cross talk [70].

In another research study, an extremely sensitive sensor to detect NH_3 gas was developed by a simple process. The working mechanism of the sensor is based on the electric resistance change of GAgs on exposing the sensor to NH_3 gas. Initially, 3D graphene hydrogels were developed in the presence and absence of different concentrations of thiourea. The resulting material on heating was converted to aerogel and finally used to detect NH$_3$ gas. The aerogels developed in the presence of thiourea were highly porous and displayed better sensitivity and selectivity towards NH$_3$ gas as compared to aerogels developed in the absence of thiourea. The sensing efficiency of aerogels was affected by the amount of thiourea used [71].

Another flexible pressure sensor based on lightweight CdS nanocomposite aerogels was prepared. The functionalized graphene sheet with a thin film of CdS was deposited via magnetron sputtering. The different analyses confirmed the deposition of nanocrystalline CdS film uniformly with 3.4% chemical composition deviation. The aforesaid nanocomposites in pressure ranging 1-5 atm demonstrated a higher response of piezoresistive and a sensitivity of 3.2 x 10^{-4} kPa^{-1} [72].

3.6   Strain sensor

The nanocomposite aerogel of bacterial cellulose/reduced graphene oxide (RGO) was developed as a potent strain sensor. This novel electrically conductive nanocomposite aerogel synthesized and dried by a supercritical drying process. To forecast the aerogels' electrical conductivity, modifications have been done on the theory of self-consistent effective medium and found that below the percolation threshold, porosity differently affects the aerogel's overall electrical conductivity. The latter result was because of the development, beyond the percolation threshold of a continuous pathway consistent with the effect of electron tunneling. It was also found that RGO in aerogels increases SSA to 252 m^{2}g^{-1}, Young's modulus to 2020 MPa, and density to 0.025 gcm^{-3}. In addition, the value of the gauge factor, i.e., 19 for nanocomposite aerogel of bacterial cellulose/RGO, confirmed its potential as a strain sensor [73].

3.7   Stress sensors

In the field of sensing and electroanalysis, fabricated graphene-based aerogels were proposed. These GAgs were fabricated with AuNPs via the freeze-drying method in a mild environment. The obtained composite aerogel was highly porous and characterized by various techniques such as X-ray photoelectron spectroscopy (XPS), Raman spectroscopy, X-ray diffraction (XRD), and scanning electron microscopy (SEM). The piece of aerogel could an electrochemical sensor directly and differs from the traditional one. The aerogel electrode showed satisfactory efficiency in hydrogen peroxide (H$_2$O$_2$) detection with the speedy response, low limit of detection, and high reproducibility. This research suggested an innovative approach for graphene relevancies in sensing and electroanalysis [74].

3.8   Hydrogen peroxide sensor

Other Ni_3N NPs based 3D graphene aerogel was developed to detect H_2O_2 and enzyme-free glucose ($C_6H_{12}O_6$). The processes involved in the synthesis were a hydrothermal reaction, freeze-drying process, and lastly, calcined in NH_3 environment. The resulted 3D Ni_3N NPs/graphene aerogel composites possessed outstanding electrochemical efficiency for the $C_6H_{12}O_6$ oxidation and H_2O_2 reduction with higher catalytic rate constant. The obtained sensor provided a low limit of detection i.e., 0.04 and 1.80 mM, a broad range of detection i.e., 0.1 mM, 7645.3 mM and 5 mM, 75.13 mM, speedy response time within 3 and 5 seconds and high sensitivity i.e., 905.6 and 101.9 $mAmM^{-1}$ cm^{-2} for the detection of $C_6H_{12}O_6$ and H_2O_2, respectively. In addition, this $C_6H_{12}O_6$ and H_2O_2 sensors displayed adequate reproducibility, the stability of long-term storage, and selectivity. In human blood serum, these sensors can effectively detect $C_6H_{12}O_6$ and H_2O_2 as well as beneficial for electrocatalytic applications [75]

3.9   Electrochemical sensor

In the case of biomedical areas, there are several biomedical devices where the electrochemical devices are utilized. For example, in the case of cancer disease, early-stage treatment is very critical to avoid the problem of metastasis. In such cases, the electrochemical sensors play a very important role. These sensors i.e. electrochemical sensors, offer detection of a cancer cell on the basis of octadecylamine (OA)-functionalized graphene aerogel microspheres (FA-GAM-OA) and folic acid (FA) detection. The FA-GAM-OA provides high electronic conductivity (2978.2 Sm^{-1}) and a large surface area (1723.6 m^2g^{-1}) to the material. The ordered alignment and covalent linkage of the folic acid groups at the FA-GAM-OA surface show strong affection with specific cancer and trap these cells with high capture efficiency. The electrochemical sensor based on FA-GAM-OA offers tremendous analytical performance, and it is helpful in the detection of liver cancer cells with a linear range of 5 to 105 cell mL^{-1} with a limit of detection of 5 cells mL^{-1} (S/N = 3) [76].

Conclusions

The aerogels play an exceptional role in the development and advancement of sensors with fast response, higher sensitivity, low detection limit, and fast recovery rate and exceptional performances owing to their remarkable properties. Hence, it is concluded that the combination of 3D porous structure and low dimensional active building blocks offers the aerogel with attractive features. Consequently, the combination offers the best sensor applications as well. Though, there are some drawbacks that need to be addressed such as:

(1) The cost-effective and industrial-scale production of high-quality aerogels.

2) Development of sensors from aerogels, like gas sensors and biosensors, without affecting the porous structure.

3) The role of the structure of aerogels is very crucial in sensing performance. However, the precise structural control of the aerogel, together with the composition, size, crystallinity, defect, pore structure, and morphology, is still a great challenge.

4) The development of highly integrated sensors with multifunctional capability is still a great challenge.

Acknowledgments

Dr. Sapna Raghav is thankful to the Department of Chemistry, Banasthali Vidyapith. Dr. Pallavi Jain, is grateful SRM Institute of Science & Technology, Delhi NCR campus, Modinagar. Dr. Dinesh Kumar is thankful to DST, New Delhi, for the financial support extended. (Sanctioned vide project Sanction Order F. No. DST/TM/WTI/WIC/2K17/124(C).

References

[1] R.W. Pekala, S.T. Mayer, J.L. Kaschmitter, F.M. Kong, Attia, Y.A., Carbon aerogels: an update on structure, properties, and applications. In Sol−gel processing and applications; Ed.; Plenum Press: New York, (1994) p 369. https://doi.org/10.1007/978-1-4615-2570-7_32

[2] Y. Hanzawa, K. Kaneko, R.W. Pekala, M.S. Dresselhaus, Activated carbon aerogels, Langmuir 12 (1996) 6167-6169. https://doi.org/10.1021/la960481t

[3] S.S. Kistler, Coherent expanded aerogels, J. Phys. Chem. 36 (1932) 52-60-64. https://doi.org/10.1021/j150331a003

[4] A. García, F. Carrillo, J. Oliva, T. Esquivel, S. Díaz, Effects of Eu content on the luminescent properties of Y_2O_3:Eu^{3+} aerogels and $Y(OH)_3$/Y_2O_3:Eu^{3+}@SiO_2 glassy aerogels, Ceram. Int. 43 (2017) 12196-12204. https://doi.org/10.1016/j.ceramint.2017.06.079

[5] S.S. Kistler, Coherent expanded aerogels and jellies, Nature 127 (1931) 741. https://doi.org/10.1038/127741a0

Materials Research Forum LLC
https://doi.org/10.21741/9781644901298-8

[6] N. Hüsing, U. Schubert, Aerogels-airy materials: chemistry, structure, and properties, Angew. Chem., Int. Ed. 1998, 37, 22. https://doi.org/10.1002/(SICI)1521-3773(19980202)37:1/2<22::AID-ANIE22>3.0.CO;2-I.

[7] S.B. Riffat, G. Qiu, A review of state-of-the-art aerogel applications in buildings, Int. J. Low-Carbon Technol. 8 (2013) 1-6 https://doi.org/10.1093/ijlct/cts001

[8] S.M. Jones, Aerogel: space exploration applications, J. Sol-Gel Sci. Technol. 2006, 40, 351-357. https://doi.org/10.1007/s10971-006-7762-7.

[9] S.M. Jones, A method for producing gradient density aerogel, J. Sol-Gel Sci. Technol. 2007, 44, 255-258. https://doi.org/10.1007/s10971-007-1618-7

[10] Y. Lee, J.W. Choi, D.J. Suh, J.M. Ha, C.H. Lee, Ketonization of hexanoic acid to diesel-blendable 6-undecanone on the stable zirconia aerogel catalyst, Appl. Catal. A-Gen. 506 (2015) 288-293. https://doi.org/10.1016/j.apcata.2015.09.008

[11] M.T. Noman, M.A. Ashraf, A. Ali, Synthesis and applications of nano-TiO_2: a review, Environ. Sci. Pollut. Res. 26 (2019) 3262-3291. https://doi.org/10.1007/s11356-018-3884-z

[12] X. Peng, X. Zhang, L. Wang, L. Hu, S.H.S. Cheng, C. Huang, B. Gao, F. Ma, K. Huo, P. K. Chu, Vanadium carbide nanoparticles encapsulated in graphitic carbon network nanosheets: A high-efficiency electrocatalyst for hydrogen evolution reaction, Nano Energy. 26 (2016) 603-609. https://doi.org/10.1016/j.nanoen.2016.06.020

[13] F. Maroni, A. Birrozzi, G. Carbonari, F. Croce, R. Tossici, S. Passerini, F. Nobili, Graphene/V_2O_5 cryogelcomposite as a high-energy cathode material for lithium-ion batteries,, Chem. Electro. Chem. 26 (2017) 613-619. https://doi.org/10.1002/celc.201600798

[14] Q. An, Y. Li, H. DeogYoo, S. Chen, Q. Ru, L. Mai, Y. Yao, Graphene decorated vanadium oxide nanowire aerogel for long-cycle-life magnesium battery cathodes, Nano Energy 18 (2015) 265-272. https://doi.org/10.1016/j.nanoen.2015.10.029

[15] N. Leventis, N. Chandrasekaran, A. G. Sadekar, S. Mulik, C. Sotiriou-Leventis, The effect of compactness on the carbothermal conversion of interpenetrating metal oxide/resorcinol-formaldehyde nanoparticle networks to porous metals and carbides, J. Mater. Chem. 20 (2010) 7456-7471. https://doi.org/10.1039/C0JM00856G.

[16] N.C. Bigall, A.K. Herrmann, M. Vogel, M. Rose, P. Simon, W. Carrillo-Cabrera, D. Dorfs, S. Kaskel, N. Gaponik, A. Eychmüller, Hydrogels and aerogels from noble metal nanoparticles, Angew. Chem. Int. Ed. 48 (2009) 9731-9734. https://doi.org/10.1002/anie.200902543

[17] W. Liu, A.K. Herrmann, D. Geiger, L. Borchardt, F. Simon, S. Kaskel, N. Gaponik, A. Eychmüller, High performance electrocatalysis on palladium aerogels, Angew. Chem. Int. Ed. 51 (2012) 5743-5747. https://doi.org/10.1002/anie.201108575

[18] A. Freytag, S. Sanchez-Paradinas, S. Naskar, N. Wendt, M. Colombo, G. Pugliese, J. Poppe, C. Demirci, I. Kretschmer, D.W. Bahnemann, P. Behrens,Versatile aerogel fabrication by freezing and subsequent freeze-drying of colloidal nanoparticle solutions, Angew. Chem. Int. Ed. 55 (2016) 1200-1203. https://doi.org/10.1002/anie.201508972

[19] W. Liu, P. Rodriguez, L. Borchardt, A. Foelske, J. Yuan, A.K. Herrmann, D. Geiger, Z. Zheng, S. Kaskel, N. Gaponik, R. Kotz, T.J. Schmidt, A. Eychmüller, High performance electrocatalysts for the oxygen reduction reaction, Angew. Chem. Int. Ed. 52 (2013) 9849-9852. https://doi.org/10.1002/anie.201303109

[20] C. Zhu, D. Wen, M. Oschatz, M. Holzschuh, W. Liu, A.K. Herrmann, F. Simon, S. Kaskel, A. Eychmüller, Kinetically controlled synthesis of PdNi bimetallic porous nanostructures with enhanced electrocatalytic activity, Small 11 (2015) 1430-1434. https://doi.org/10.1002/smll.201401432

[21] A.S. Douk, M. Farsadrooh, F. Damanigol, A.A. Moghaddam, H. Saravani, M. Noroozifar, Porous three-dimensional network of Pd–Cu aerogel toward formic acid oxidation, RSC Adv. 8 (2018) 23539-23545. https://doi.org/10.1039/C8RA03718C

[22] D. Wen, W. Liu, D. Haubold, C. Zhu, M. Oschatz, M. Holzschuh, A. Wolf, F. Simon, S. Kaskel, A. Eychmüller, Gold aerogels: three-dimensional assembly of nanoparticles and their use as electrocatalytic interfaces, ACS Nano 10 (2016) 2559-2567. https://doi.org/10.1021/acsnano.5b07505

[23] F. Qian, P.C. Lan, M.C. Freyman, W. Chen, T. Kou, T.Y. Olson, C. Zhu, M.A. Worsley, E.B. Duoss, C.M. Spadaccini, Qian, Ultralight conductive silver nanowire aerogels, Nano Lett. 17 (2017) 7171-7176. https://doi.org/10.1021/acs.nanolett.7b02790

[24] L. Lu, X.F. Sun, J. Ma, D.X. Yang, H.H. Wu, B.X. Zhang, J.L. Zhang, B.X. Han, Highly efficient electroreduction of CO_2 to methanol on palladium–copper bimetallic

Materials Research Forum LLC
https://doi.org/10.21741/9781644901298-8

aerogels, Angew. Chem. Int. Ed. 57 (2018) 14149-14349.
https://doi.org/10.1002/ange.201808964-14349

[25] Z.L. Yu, B. Qin, Z.Y. Ma, J. Huang, S.C. Li, H.Y. Zhao, H. Li, Y.B. Zhu, H.A. Wu, S.H. Yu, Superelastic hard carbon nanofiber aerogels, Adv. Mater. 31 (2019) 1900651 https://doi.org/10.1002/adma.201900651

[26] S.C. Li, B.C. Hu, Y.W. Ding, H.W. Liang, C. Li, Z.Y. Yu, Z.Y. Wu, W.S. Chen, S.H. Yu, Wood-derived ultrathin carbon nanofiber aerogels, Angew. Chem., Int. Ed. 57 (2018) 7085-7090. https://doi.org/10.1002/anie.201802753

[27] P. Hao, Z.H. Zhao, Y.H. Leng, J. Tian, Y.H. Sang, R.I. Boughton, C.P. Wong, H. Liu, B. Yang,Graphene-based nitrogen self-doped hierarchical porous carbon aerogels derived from chitosan for high performance supercapacitors, Nano. Energy 15 (2015) 9-23. https://doi.org/10.1016/j.nanoen.2015.02.035

[28] Y. Xu, K. Sheng, C. Li, G. Shi, Self-assembled graphene hydrogel via a one-step hydrothermal process, ACS Nano 4 (2010) 4324-4330. https://doi.org/10.1021/nn101187z

[29] F. Guo, Y. Jiang, Z. Xu, Y. Xiao, B. Fang, Y. Liu, W. Gao, P. Zhao, H. Wang, C. Gao, Highly stretchable carbon aerogels, Nat. Commun. 9 (2018) 881. https://doi.org/10.1038/s41467-018-03268-y

[30] M. Salzano de Luna, Y. Wang, T. Zhai, L. Verdolotti, G.G. Buonocore, M. Lavorgna, H. Xia, Nanocomposite polymeric materials with 3D graphene-based architectures: from design strategies to tailored properties and potential applications, Prog. Polym. Sci. 89 (2019) 213-249. https://doi.org/10.1016/j.progpolymsci.2018.11.002

[31] H. Zhuo, Y. Hu, X. Tong, Z. Chen, L. Zhong, H. Lai, L. Liu, S. Jing, Q. Liu, C. Liu, X. Peng, R. Sun, A super compressible, elastic, and bendable carbon aerogel with ultrasensitive detection limits for compression strain, pressure, and bending angle, Adv. Mater. 30 (2018) 1706705. https://doi.org/10.1002/adma.201706705

[32] T. Gacoin, L. Malier, J.P. Boilot, New transparent chalcogenide materials using a sol-gel process, Chem. Mater. 9 (1997) 1502-1504. https://doi.org/10.1021/cm970103p

[33] V. Sayevich, B. Cai, A. Benad, D. Haubold, L. Sonntag, N. Gaponik, V. Lesnyak, A. Eychmüller, 3D assembly of all-inorganic colloidal nanocrystals into gels and

aerogels, Angew. Chem. Int. Ed. 55 (2016) 6334-6338.
https://doi.org/10.1002/anie.201600094

[34]   S. Naskar, J.F. Miethe, S. Sánchez-Paradinas, N. Schmidt, K. Kanthasamy, P. Behrens, H. Pfnür, N.C. Bigall, Photoluminescent aerogels from quantum wells, Chem. Mater. 28 (2016) 2089-2099. https://doi.org/10.1021/acs.chemmater.5b04872

[35]   A. Du, B. Zhou, J.Y. Gui, G.W. Liu, Y.N. Li, G.M. Wu, J. Shen, Z.H. Zhang, Thermal and mechanical properties of density-gradient aerogels for outer-space hypervelocity particle capture, Acta Phys.-Chim. Sin. 28 (2012) 1189-1196. https://doi.org/10.3866/PKU.WHXB201202292

[36]   H. Nawaz, P.A.R. Pires, O.A. El Seoud, Kinetics and mechanism of imidazole-catalyzed acylation of cellulose in LiCl/N, N-dimethylacetamide, Carbohydr. Polym. 92 (2013) 997-1005. https://doi.org/10.1016/j.carbpol.2012.10.009

[37]   T. Hongo, M. Inamoto, M. Iwata, T. Matsui, K. Okajima, Morphological and structural formation of the regenerated cellulose membranes recovered from its cuprammonium solution using aqueous sulfuric acid, J. Appl. Polym. Sci. 72 (1999) 1669-1678. https://doi.org/10.1002/(SICI)1097-4628(19990624)72:13<1669::AID-APP3>3.0.CO;2-L

[38]   D.A. Osorio, B.E.J. Lee, J.M. Kwiecien, X. Wang, I. Shahid, A.L. Hurley, E.D. Cranston, K. Grandfield, Cross-linked cellulose nanocrystal aerogels as viable bone tissue scaffolds, Acta Biomater. 87 (2019) 152-165. https://doi.org/10.1016/j.actbio.2019.01.049

[39]   C.A. García-González, M. Alnaief, I. Smirnova, Polysaccharide-based aerogels-Promising biodegradable carriers for drug delivery systems, Carbohydr. Polym. 86 (2011) 1425-1438. https://doi.org/10.1016/j.carbpol.2011.06.066

[40]   J. Wang, X. Wang, X. Zhang, Cyclic molecule aerogels: a robust cyclodextrin monolith with hierarchically porous structures for removal of micropollutants from water, J. Mater. Chem. A 5 (2017) 4308-4313. https://doi.org/10.1039/C6TA09677H

[41]   Y. Liu, Y. Su, J. Guan, J. Cao, R. Zhang, M. He, Z. Jiang, Asymmetric aerogel membranes with ultrafast water permeation for the separation of oil-in-water emulsion, ACS Appl. Mater. Interfaces 10 (2018) 26546-26554. https://doi.org/10.1021/acsami.8b09362

[42]   G. Zu, T. Shimizu, K. Kanamori, Y. Zhu, A. Maeno, H. Kaji, J. Shen, K. Nakanishi, Transparent, super flexible doubly cross-linked polyvinyl polymethyl

siloxane aerogel super insulators via ambient pressure drying, ACS Nano 12 (2018) 521-532. https://doi.org/10.1021/acsnano.7b07117

[43] H. Long, A. Harley-Trochimczyk, T. Pham, Z. Tang, Tielin Shi, A. Zettl, C. Carraro, M. A. Worsley, R. Maboudian, High surface area MoS_2/graphene hybrid aerogel for ultrasensitive NO_2 detection, Adv. Funct. Mater. 26 (2016) 5158–5165. https://doi.org/10.1002/adfm.201601562

[44] L. Li, M. Liu, S. He, W. Chen, Freestanding 3D mesoporous Co_3O_4 @ carbon foam nanostructures for ethanol gas sensing, Anal. Chem. 86 (2014) 7996-8002. https://doi.org/10.1021/ac5021613

[45] S. Dolai, S.K. Bhunia, R. Jelinek, Carbon-dot-aerogel sensor for aromatic volatile organic compounds, Sens. Actuators B 241 (2017) 607. https://doi.org/10.1016/j.snb.2016.10.124

[46] J. Wu, Z. Li, X. Xie, K. Tao, C. Liu, K.A. Khor, J. Miao, L.K. Norford, 3D super hydrophobic reduced graphene oxide for activated NO_2 sensing with enhanced immunity to humidity, J. Mater. Chem. A 6 (2018) 478-488. https://doi.org/10.1039/C7TA08775F

[47] D.L. Plata, Y.J. Briones, R.L. Wolfe, M.K. Carroll, S.D. Bakrania, S.G. Mandel, A.M. Anderson, Aerogel-platform optical sensor for oxygen gas, J. Non-Cryst. Solids 350 (2004) 326-335. https://doi.org/10.1016/j.jnoncrysol.2004.06.046

[48] J.T. Korhonen, P. Hiekkataipale, J. Malm, M. Karppinen, O. Ikkala, R.H.A. Ras, Inorganic hollow nanotube aerogels by atomic layer deposition onto native nanocellulose templates, ACS Nano 5 (2011) 1967-1974. https://doi.org/10.1021/nn200108s

[49] F. Yang, J. Zhu, X. Zou, X. Pang, R. Yang, S. Chen, Y. Fang, T. Shao, X. Luo, L. Zhang, Three-dimensional TiO_2/SiO_2 composite aerogel films via atomic layer deposition with enhanced H_2S gas sensing performance, Ceram. Int. 44 (2018) 1078-1085. https://doi.org/10.1016/j.ceramint.2017.10.052

[50] T.T. Li, R.R. Zheng, H. Yu, Y. Yang, T.T. Wang, X.T. Dong, Versatile aerogels for sensors, RSC Adv. 7 (2017) 39334. https://doi.org/10.1002/smll.201902826

[51] E. Barrios, D. Fox, Y.Y.L.Sip, R.Catarata, J.E. Calderon,N. Azim,S. Afrin, Z. Zhang, L. Zha, Nanomaterials in advanced, high-performance aerogel composites: a review, Polymers (Basel) 11 (2019) 726. https://doi.org/10.3390/polym11040726

[52]   A. Walcarius, M.M. Collinson, Analytical chemistry with silica sol-gels: traditional routes to new materials for chemical analysis, Annu. Rev. Anal. Chem. 2 (2009) 121-143. https://doi.org/10.1146/annurev-anchem-060908-155139

[53]   M.K. Carroll, A.M. Anderson, In: M.A. Aegerter, N. Leventis, M.M. Koebel (Eds.), Aerogels handbook, Springer, New York, 2011, pp. 637-650 https://doi.org/10.1007/978-1-4419-7589-8_27

[54]   T. Wagner, S. Haffer, C. Weinberger, D. Klaus, M. Tiemann, Mesoporous materials as gas sensors, Chem. Soc. Rev. 42 (2013) 4036-4053 https://doi.org/10.1039/C2CS35379B

[55]   R. Wang, G. Li, Y. Dong, Y. Chi, and G. Chen, Carbon quantum dot-functionalized aerogels for NO_2 gas sensing, Anal. Chem. 85 (2013) 8065-8069 https://doi.org/10.1021/ac401880h

[56]   T. Wagner, S. Haffer, C. Weinberger, D. Klaus, M. Tiemann, Mesoporous materials as gas sensors, Chem. Soc. Rev. 42 (2013) 4036-4053. https://doi.org/10.1039/C2CS35379B

[57]   M. Barczak, C. McDonagh, D. Wencel, Micro- and nanostructured sol-gel-based materials for optical chemical sensing (2005–2015), Microchim. Acta 183 (2016) 2085-2109. https://doi.org/10.1007/s00604-016-1863-y

[58]   C.T. Wang, C.L. Wu, I.C. Chen, Y.H. Huang, Humidity sensors based on silica nanoparticle aerogel thin films, Sens. Actuators B 107 (2005) 402–410. https://doi.org/10.1016/j.snb.2004.10.034

[59]   J.E. Amonette, J. Matyas, Functionalized silica aerogels for gas-phase purification, sensing, and catalysis: A review, Micropor. Mesopor. Mat. 250 (2017) 100-119. https://doi.org/10.1016/j.micromeso.2017.04.055

[60]   C.T. Wang, C.L. Wu, Electrical sensing properties of silica aerogel thin films to humidity, Thin Solid Films 496 (2006) 658-664. https://doi.org/10.1016/j.tsf.2005.09.001

[61]   M.R. Ayers, A.J. Hunt, Molecular oxygen sensors based on photoluminescent silica aerogels, J Non-Cryst. Solids 225 (1998) 343-347. https://doi.org/10.1016/S0022-3093(98)00051-9

[62] N. Leventis, I.A. Elder, D.R. Rolison, M.L. Anderson, C.I. Merzbacher, Silica nano architectures incorporating self-organized protein superstructures with gas-phase bioactivity, Chem. Mater. 11 (1999) 2837-2845. https://doi.org/10.1021/nl034646b

[63] N. Leventis, A.M.M. Rawashdeh, I.A. Elder, J. Yang, A. Dass, C. Sotiriou-Leventis, Synthesis and characterization of Ru(II) tris(1,10-phenanthroline)-electron acceptor dyads incorporating the 4-benzoyl-N-methylpyridinium cation or N-benzyl-N'-methyl viologen. Improving the dynamic range, sensitivity, and response time of sol−gel-based optical oxygen sensors, Chem. Mater. 16 (2004) 1493-1506. https://doi.org/10.1021/cm034999b

[64] X. Xu, R. Wang, P. Nie, Y. Cheng, X. Lu, L. Shi, J. Sun, Copper nanowire-based aerogel with tunablepore structure and its application as flexible pressure sensor, ACS Appl. Mater. Interface, 9 (2017) 14273−14280. https://doi.org/10.1021/acsami.7b02087

[65] X. Chen, H. Liu, Y. Zheng, Y. Zhai, X. Liu, C. Liu, L. Mi, Z. Guo, C. Shen, Highly compressible and robust polyimide/carbon nanotube composite aerogel for high-performance wearable pressure sensor, ACS Appl. Mater. Interface, 11 (2019) 42594-42606. https://doi.org/10.1021/acsami.9b14688

[66] J. Kehrle, T.K. Purkait, S. Kaiser, K.N. Raftopoulos, M. Winnacker, T. Ludwig, M. Aghajamali, M. Hanzlik, K. Rodewald, T. Helbich, Christine M. Papadakis, J.G.C. Veinot, B. Rieger, Superhydrophobic silicon nanocrystal−silica aerogel hybrid materials: synthesis, properties, and sensing application, Langmuir, 34 (2018) 4888-4896. https://doi.org/10.1021/acs.langmuir.7b03746

[67] Q. Luo, H. Zheng, Y. Hu, H. Zhuo, Z. Chen, X. Peng, L. Zhong, Carbon nanotube/chitosan-based elastic carbon aerogel for pressure sensing, industrial and engineering chemical research, 58 (2019) 17768−17775. https://doi.org/10.1021/acs.iecr.9b02847

[68] I. Ali, L. Chen, Y. Huang, L. Song, X. Lu, B. Liu, L. Zhang, J. Zhang, L. Hou, T. Chen, Humidity-responsive gold aerogel for real-time monitoring of human breath, Langmuir 34 (2018) 4908−4913. https://doi.org/10.1021/acs.langmuir.8b00472

[69] H. Zhuo, Y. Hu, Z. Chen, X. Peng, L. Liu, Q. Luo, J. Yi, C. Liu, L. Zhong, A carbon aerogel with super mechanical and sensing performances for wearable piezoresistive sensors, J. Mater. Chem. 7 (2019) 8092-8100. https://doi.org/10.1039/C9TA00596J

[70]   S. Han, F. Jiao, Z.U. Khan, J. Edberg, S. Fabiano, X. Crispin, Thermoelectric polymer aerogels for ressure-temperature sensing applications, Adv. Funct. Mater. 27 (2017) 1703549. https://doi.org/10.1002/adfm.201703549

[71]   T. Alizadeh, F. Ahmadian, Thiourea-treated graphene aerogel as a highly selective gas sensor for sensing of trace level of ammonia, Anal. Chim. Acta897 (2015) 87-95. https://doi.org/10.1016/j.aca.2015.09.031

[72]   I. Plesco, M. Dragoman, J. Strobel, L. Ghimpu, F. Schütt, A. Dinescu, V. Ursaki, L. Kienle, R. Adelung, I. Tiginyanu, Flexible pressure sensor based on graphene aerogel microstructures functionalized with CdS nanocrystalline thin film, Superlattice Microst. 117 (2018) 418-422 https://doi.org/10.1016/j.spmi.2018.03.064

[73]   H. Hosseinia, M. Kokabi, S. Mohammad Mousavi, BC/rGO conductive nanocomposite aerogel as a strain sensor, Polymer 137 (2017) 82-96. https://doi.org/10.1016/j.polymer.2017.12.068

[74]   X. Houa, R. Zhanga, D. Fang, Super elastic, fatigue resistant and heat insulated carbon nanofiber aerogels for piezo resistive stress sensors, Ceramics Inter. 46 (2020) 2122-2127. https://doi.org/10.1016/j.ceramint.2019.09.195

[75]   D. Yin, X. Bo, J. Liu, L. Guo, A novel enzyme-free glucose and H_2O_2 sensor based on 3D graphene aerogels decorated with Ni_3N nanoparticles, Anal. Chim. Acta 1038 (2018) 1-10. https://doi.org/10.1016/j.aca.2018.06.086

[76]   L. Ruiyi, C. Fangchao, Z. Haiyan, S. Xiulan, L. Zaijun, Electrochemical sensor for detection of cancer cell based on folic acid and octadecylamine-functionalized graphene aerogel microspheres, Biosens. Bioelectron. 119 (2018)156-162. https://doi.org/ 10.1016/j.bios.2018.07.060

Aerogels II: Preparation, Properties and Applications   Materials Research Forum LLC
Materials Research Foundations **97** (2021) 168-182   https://doi.org/10.21741/9781644901298-9

Chapter 9

Aerogels as Pesticides

Sajjad Ali[1*], Waseem Akram[1], Asif Sajjad[1], Qaiser Shakeel[2], Muhammad Irfan Ullah[3]

[1]Department of Entomology, The Islamia University of Bahawalpur, Bahawalpur 63100, Pakistan

[2]Department of Plant Pathology, The Islamia University of Bahawalpur, Bahawalpur 63100, Pakistan

[3]Department of Entomology, College of Agriculture, University of Sargodha, Sargodha 40100, Pakistan

*sajjad.ali@iub.edu.pk

Abstract

Aerogels, composed of complex network of interlinked nanostructures, show 50% non-solid volume. Due to their unique properties, they are used as a carrier for active ingredients used to control agricultural pests as well as veterinary medicines. They can also be used as a carrier material for the application of entomopathogenic bacteria and viruses for the biological control of pests. Many aerogel-based formulations of herbicides, insecticides, acaricides, fungicides, bactericides, rodenticides, nematicides, piscicides and molluscicides effectively control the target pests. Practically, aerogels enhance the effectiveness of insecticides by increasing their penetrations. Furthermore, intensive research is required to develop latest aerogel-based pesticides with better utilization under effective integrated pest management programs in agriculture.

Keywords

Aerogels, Pesticides, Integrated Pest Management, Formulations, Pest Control

Contents

Materials Research Forum LLC
https://doi.org/10.21741/9781644901298-9

1.   Introduction

Aerogels are usually, porous and complex mix of nanostructures with solid and non-solid parts with 50:50 ratio, approximately [1]. Aerogels are made up of a wide range of substances i.e. silica, carbon, biological polymers (pectin, gelatin and agar), organic polymer (epoxies, polystyrenes, poly acrylates, polyurethanes, phenol-formaldehyde and resorcinol-formaldehyde), metal oxides (iron oxide and praseodymium oxide), semiconductors, metals (gold and copper) and carbon nano-tubes [2]. Due to dual structural features i.e. nano-scale skeleton and condensed state matter and also the higher porosity, aerogels exhibit unique characters i.e. specific refractive index, sonic velocity, surface area, dielectric constant, adjustable density and low thermal conductivity. Aerogel density may range from 1 to 1000 kg/m^3 which stimulates the notable changes in the characteristics of gel [3].

The aerogels preparation consists of these three steps: 1) Sol-gel transition (gelation), 2) Network perfection (aging), and 3) Gel-aerogel transition (drying) [4]. The microstructure of aerogels mainly depends on these three steps. Different commonly available aerogels such as oxides-based (Zirconia, Quarz and Zirkonia), Carbon-based (Pyrolysed-polymere), Polymer-based (Resorcin-Formaldehyde and Melamin-Formaldehyde), cellulosic aerogels, alginates and starch-based aerogels [3]. Silica aerogels are the most widely used because of their low density and thermal conductivity (0.01 W/m.k), hydrophobicity, high capability of sound absorption and low velocity (100 m/s), low refractive index (1.05) and dielectric constant (1-2), higher porosity (90-99 %), higher optical transparency (99 %) and surface area (1000 m^2/g) [2, 5-9]. These unique characters of aerogels are responsible for their wide range of applications (Table 1) [10-26].

Table 1. *Various applications of aerogels [10-26].*

S. No.	Applications	References
1	Aerogels as composite	[10, 11]
2	Aerogels as an absorbent	[12, 13]
3	Aerogel as a sensor	[14]
4	Aerogel used as a material with low dielectric constant	[15-17]
5	Aerogels as catalysts	[18-20]
6	Aerogel as storage media	[20-23]
7	Aerogel as template	[24]
8	Aerogel usage in agriculture	[25, 26]

2. Uses in agriculture

Aerogels serve as a carrier for active ingredients in various formulations of piscicides being used in agricultural systems for food and fiber production. The aerogels mixed as carrier medium with active ingredients are then directly applied on field crops and livestock animals. Aerogels can also be used as a carrier material for biopesticide formulations based on bacteria (*Bacillus thuringiensis*) and viruses for the biological control of pests [26].

3. Aerogels as acaricides

Ticks have developed resistance against various acaricides due to the extensive, repeated use of high dosage of acaricides [27-29]. Therefore, aerogels based acaricides are the best alternative to them. Different aerogels based acaricides are used in the form of dust to control mites, fleas and ticks (Table 2) [30-40] which are very effective against these pests. *Ixodes scapularis* has successfully been managed by DRi-Die, drione, diatomaceous earth and Safer's (insecticidal soap). Mortality percentage was very high after the application of aerogels. *Amblyomma americanum* larvae and nymph were managed through drione with 100% mortality rate [30, 31]. Egg production as well as hatching is also prevented or extremely reduced by the application of drione. Dri-Die 67 caused sudden death by desiccation in fleas (*Ctenocephalides felis* and *Xenopsylla cheopis*), mites (*Haemokzekzps glasgowi* and *Tyrophagus* sp.) and ticks (*Rhipicephalus sanguineus* and *Otobius megnini*) [32]. Lethal effect of CimeXa an insecticidal dust, against *A. americanum* and *A. maculatum* has also been reported [31]. In comparison with Surround, CimeXa was highly effective against larvae of *A. americanum* and *A. maculatum* and was found more lethal to larvae than nymphs. The larvae and nymphs showed high mortality when crawled across the layer of CimeXa than the Surround. Moreover, larvae showed almost 80% mortality after 48 hours when crawled on dried aqueous suspensions of CimeXa and Surround. A similar study was also conducted by Showler and Harlien [33] and reported that larvae and nymphs of *A. americanum* showed 90and 70% mortality respectively, after 24 hours when released on the treated skin of calves with CimeXa.

Dryacide is coated with silica aerogel which cause inhalation hazards in mites [34]. *Acarus siro* and *Tyrophagu sputrescentiae* exhibited more than 99% mortality at 75% relative humidity, 15°C temperature and Dryacide dose of 1 to 5 g/kg whereas, *Glycyphagus destructor* showed least susceptibility at dose of 3 to 5 g/kg [35]. Schulz et al. [36] measured the mortality of mites in the form of LT_{50} (Mean lethal time). The LT_{50} of liquid formulation silica-gels ranging from 5.5-12.7 hours overlapped with powdery formulation ranging from 5.1-18.7 hours.

Table 2 *Types of commercially available silica aerogel-based acaricides and their targeted pests [30-40].*

Sr. No.	Silica aerogel-based acaricides	Targeted organisms	References
1	Drione	*Ixodes scapularis* and *A. americanum*	[30, 31]
2	CimeXa	*Amblyomma americanum* and *A. maculatum*	[31, 33]
3	Dryacide	*Acarus siro, Tyrophagus putrescentiae* and *Glycyphagus destructor*	[34, 35]
4	Dri-Die 67® (SG-67)	*Ixodes scapularis, Ctenocephazides canis, C. jelis, Rhipicephalus sanguineus, Ophionyssus natricis, Ornithonyssus bacoti, Haemokzekzps glasgowi* and *Tyrophagus* sp.	[30, 32, 37-40]

4. Aerogels as insecticides

Many researchers have documented that few finely divided particles used as carrier in insecticide formulations enhanced the effectiveness of insecticides. It is determined that most of the carriers have the water sorptive and abrasive properties. These carriers comprised finely divided silicas i.e. silica aerogel and silica gel [41]. Silica aerogels are used solely in the form of insecticidal dust [42-44]. These finely divided dust particles kill insects by blocking their spiracles, absorbing the water from cuticle, being ingested and lodged between cuticle segments to prevent their movement [42, 45-48]. The only mechanism of aerogel powders is abrasion or absorption of epicuticular wax layer which causes loss of insect body fluid and ultimately death [42-44, 49-51]. To reduce the dustiness of silica aerogel without changing the characteristics, aerogels are coated with ammonium fluosilicate (5%) and35-70% glycol i.e. lower alkylene glycols [41]. The ammonium fluosilicate also increases the effectiveness of insecticidal dust without changing the ability of the silica aerogel particles to adsorb the lipids from the epicuticle [32]. Silica aerogels are pesticide free, non-toxic to mammals, most effective against pesticide resistant species and stable at low and high temperatures [52, 53]. Various types of silica aerogels-based insecticides (Table 3) [32, 33, 41, 50, 61-80] are used for the control of ants, silverfish, fleas, mosquitoes, cockroaches, crickets, bedbugs, thrips, spiders, house flies, wasps and stored grain insect pests in stores, homes, hotels, restaurants, theaters, schools, hospitals and other places where food and human beings are present [33, 44, 50, 54-58]. SG-68 is very light and fluffy silica aerogel as well as SG-67 which is commercially available in the trade name of Dri-Die 67® is similar to SG-68. Both silica aerogels contain outstanding insecticidal effectiveness due to their adequate pore diameter and large specific surface [59, 60]. Another most common silica aerogel-based dust is Drione which contain piperonylbutoxide (PBO) and pyrethrin and gives the combined effect of dehydration as well as nerve toxin [31].

CimeXa is a silica aerogel-based dust which causes dehydration in insects [81, 82]. It is considered very lethal to bed bugs and provides efficient control than the pyrethroid insecticides [61, 75]. The bedbugs exhibited 100% mortality within 24 hours when the dust products such as Tempo dust, Drione and Syloid 244 (Silica aerogel) were applied to mattress and hardboards. Drione caused similar mortality but somewhat slower than the Syloid 244 [75].

The effectiveness of silica aerogel dust (Dri-Die and Silikil-D) and aqueous suspensions against *Culex fatigans*, *Anopheles quadrimaculatus* and susceptible and Dichlorodiphenyltrichloroethane (DDT) resistant *Aedes aegypti* is very high @ 50 mg/ft^2. But the humidity had negative effect on the effectiveness of silica aerogel dusts [68]. Susceptible and DDT resistant *A. aegypti* strains showed equal mortality. Aqueous suspensions of silica aerogels were less than half as effective as dusts. Silikil-D and aqueous suspensions at high relative humidity (RH) exhibited more mortality against *A. aegypti* but was little effective at high dose. It has been found that silica aerogel insecticides also provide efficient control against termites. Silica aerogel containing 162 compounds were most effective by causing death of wingless dry wood termites by decrease in body weight due to water loss within 1-2 hours. It has also been reported that sorptive dusts were more efficient than abrasive materials [63].

Table 3 *Types of commercially available silica aerogel-based insecticides and their targeted insect pests [32, 33, 41, 50, 61-80].*

Sr. No.	Silica aerogel-based insecticides	Targeted organisms	Reference
1	CimeXa	*Cimex lectularius*	[61]
2	Dri-die 68® (SG-68)	*Periplaneta americana* and *Blattella germanica*	[62]
3	Dri-die 67® (SG-67)	Cockroaches, Termites, *Oryzaephilus mercator*, *Tribolium confusum*, *Xenopsylla cheopis*, *Rhipicephalus sanguineus*, *Otobius megnini*, *Huematojinus eurysternus*, *Triutomu protracta*, *Acheta assimilis*, *Anopheles quadrimaculatus*, *Culex fatigans* and *Aedes aegypti*	[32, 50, 63-72]
4	Santocel	Crawling insects	[41]
5	Cab-O-Sil	Lesser grain borer, *Sitophilus oryzae*, *Blattella germanica* and *Prostephanus truncates*	[41, 69, 73, 74]
6	Drione	Ants, cockroaches, silverfish, bed bugs, fleas, spiders, crickets, and wasps	[33, 66, 67, 75]
7	Dryacide	*Ephestia cautella*, *T. confusum*, *T. castaneum*, *Rhyzopertha dominica*, *S. oryzae* and *S. granaries*	[76-78]
8	Gasil 23D	*P. truncates*	[73]
9	Sipernat 22	*P. truncates*	[73]
10	Neosyl TS	*P. truncates*	[73]
11	Syloid 244	Bed bugs	[75]
12	Aerosil-R972	*P. truncates* and *T. castaneum*	[73, 79]
13	Aerosil-380	*T. confusum*	[80]

Various studies have reported the efficacy of desiccant aerogel dusts against insect pests of stored grains [57, 70, 76-78, 83, 84]. The most common commercially used silica aerogel Dri-Die 67® has proved more efficient as surface treatment in both dust as well as slurry form [53]. Treated paper sacks with silica aerogel Dri-Die 67® provides more protection for long period against *Tribolium confusum* adults than the untreated sacks [85]. Mortality of Dri-Die 67 exposed *T. confusum* and *Oryzaephilus mercator* adults was greater on breadcrumbs whereas early larval instars of *T. confusum* and *O. mercator* were more susceptible than the later instars [72]. Quantity and type of food, type of adhesive, amount of silica aerogel particles and exposure time can influence the mortality [71, 72]. As exposure time increased, adults of both *T. confusum* and *O. mercator* survived only if the food was available at shorter intervals [72]. Loschiavo [70] revealed that all *O. mercator* adults died within 48 hours when walking over a thin layer of Dri-Die 67 for a period of 20-30 seconds in the absence of food whereas only 17% died when food was available. Loschiavo [71] also reported that both *T. confusum* and *O. Mercator* adults showed 100 % mortality on heavy coatings after 3 hours exposure. Gowers and Patourel [86] concluded that loose dust deposits of silica suspensions were most toxic on glass surface while dried dust deposits were most toxic on vinyl surface to *Sitophilus granaries*. Vrba et al. [80] checked the efficacy of Aerosil-380 against *T. confusum* at varying exposure time and reported that mortality is initiated due to the scarcity of food and exposure to the Aerosil-380. Le-Patourel and Singh [79] measured the toxicity of Aerosil R972 and Cab-O-Sil M5 against *T. castaneum* as a carrier to prepare permethrin, deltamethrin and cypermethrin formulations. They found that intermediate and low concentrations of these pyrethroids greatly reduced the 48 h LC_{50} of both silica aerogels whereas high concentrations antagonized their toxic action because of knockdown effects. Desmarchelier and Dines [76] conducted a study to check the efficacy of Dryacide on wheat against *T. castaneum*, *S. oryzae*, *S. garnarius*, *R.dominica* and *Ephestia cautella*. Dryacide was found most effective at both immature and adult stages of insects. They also found that when Dryacide treated wheat grains milled without cleaning, <3 % Dryacide was found in the wheat flour. Another similar study was carried out by Aldryhim [77] and concluded that *S. granarius* immatures and adults were more susceptible to Dryacide than the *T. confusum* at 30 °C and 40 % relative humidity. Dryacide was found safer because of no adverse effect on wheat flour as well as baking qualities. Silica aerogel dusts (Gasil and Aerosil) provided efficient control against lesser grain borer (*R. dominica*) and larger grain borer (*Prostephanus truncatus*) [73, 78]. The rice weevil, *S. oryzae* showed highest mortality after 21 days when exposed to Cab-O-Sil 500 and Cab-O-Sil 750 [74]. Silica aerogels also showed promise for the efficient control of dermestid beetles i.e. *Dermestes ater*, *D. frischii* and *Necrobia rufipes* [87]. The most commonly used silica aerogels against cockroaches are Dri-Die 67 and Dri-Die 68. In laboratory studies to determine the effectiveness of various dusts on household pests, the writers [64, 65, 88] discovered a silica aerogel, SG 67, to be singularly effective against cockroaches. American, German, brown banded and oriental cockroaches were found dead within 24 hours when exposed to Dri-Die 67 (SG-67) and held at 100 % R.H. [32]. Whereas another study showed different results from the previous statement. The SG-67 exposed German cockroaches were found dead after 75 minutes, oriental after 180 minutes and American cockroaches were dead after 270 minutes [65]. German and American cockroaches exhibited similar rates of mortality after ingestion of large quantities of silica aerogel, Dri-die 68® [62]. Some other studies also reported that silica aerogel-based insecticides i.e. Drione and Dri-Die 67 were efficient for the control of cockroaches [66, 67]. Laboratory tests showed that the silica gel materials like Dri-Die®, Micro-Cel-E®, Cab-O-Sil®, Micro-Cel-C, Celite® 209 and Celite 270 were more effective

Materials Research Forum LLC
https://doi.org/10.21741/9781644901298-9

to control and repel the German cockroaches than boric acid powder. Addition of 0.1% Dri-die, 1.0% Micro-Cel-C and 0.1% Cab-O-Sil to boric acid resulted in the increase in their efficacy [69]. Previous studies indicate their effective utilization to manage the insect pests. The cereal insect pests were found highly susceptible to these aerogel pesticides.

Conclusion

Various aerogels-based products possess pesticidal potential and are effective to control variety of insect and non-insect pests of economic importance in agriculture and livestock sectors. Specially, they have been proved as potential protectants for common storage insect pests of cereals and grains. Their effectiveness can be enhanced under low humid conditions to induce high mortality by causing desiccation and removal of cuticular waxy layer of the exoskeleton of exposed arthropods. Modern technology, to produce aerogel dusts and dry acids, has reduced the associated health hazards with them. Their potential use at large-scale, commercial agricultural storage and at small-scale holdings in the developing countries needs to be strengthen through aggressive research by widening rage and applications of aerogels.

References

[1] C.A. García-Gonzalez, M. Alnaief, I. Smirnova, Polysaccharide-based aerogels-promising biodegradable carriers for drug delivery systems, Carbohyd. Polym. 86 (2011) 1425-1438. https://doi.org/10.1016/j.carbpol.2011.06.066

[2] L. Ratke, Aerogels-Structure, properties and applications, Institut für Material physikim Weltraum (2006) 51147.

[3] A. Du, B. Zhou, Z.H. Zhang, J. Shen, A special material or a new state of matter: a review and reconsideration of the aerogel Materials. 6 (2013) 941-968. https://doi.org/10.3390/ma6030941

[4] A.K. Nayak, B. Das, Introduction to polymeric gels, in: K. Pal, I. Banerjee (Eds. 1st), Polymeric Gels, Woodhead Publishing. (2018) 3-27. https://doi.org/10.1016/B978-0-08-102179-8.00001-6

[5] C.A.M. Mulder, J.G. Van-Lierop, Preparation, densification and characterization of autoclave dried SiO2 gels, in: J. Fricke (Eds.), Aerogels, Springer, Berlin, Germany. (1986) 68–75. https://doi.org/10.1007/978-3-642-93313-4_8

[6] G.C. Bond, S. Flamerz, Structure and reactivity of titania-supported oxides. Part 3: reaction of isopropanol over vanadia-titania catalysts, Appl. Catal. 33(1) (1987) 219-230. https://doi.org/10.1016/S0166-9834(00)80594-1

[7] L.W. Hrubesh, Aerogels: the world's lightest solids, Chem. Ind. 24 (1990) 824–827.

[8] J. Fricke, A. Emmerling, Aerogels, preparation, properties, applications, Structure and Bonding 77: Chemistry, Spectroscopy and Applications of Sol-Gel Glasses, Springer, Berlin, Germany. (1992) 37–87. https://doi.org/10.1007/BFb0036965

[9] J.L. Gurav, I.K. Jung, H.H. Park, E.S. Kang, D.Y. Nadargi, Silica aerogel: synthesis and applications, J. Nanomater. (2010) 1-11. https://doi.org/10.1155/2010/409310

[10] L.L. Casas, A. Roig, E. Rodriguez, E. Molins, J. Tejada, J. Sort, Silica aerogel-iron oxide nanocomposites: structural and magnetic properties, J. Non-Cryst. Solids. 285(1–3) (2001) 37–43. https://doi.org/10.1016/S0022-3093(01)00429-X

[11] L.L. Casas, A. Roig, E. Molins, J.M. Greneche, J. Asenjo, J. Tejada, Iron oxide nanoparticles hosted in silica aerogels, Appl. Phys. A 74(5) (2002) 591–597. https://doi.org/10.1007/s003390100948

[12] A.V. Rao, S.D. Bhagat, H. Hirashima, G.M. Pajonk, Synthesis of flexible silica aerogels using methyltrimethoxysilane (MTMS) precursor, J. Colloid. Interf. Sci. 300(1) (2006) 279–285. https://doi.org/10.1016/j.jcis.2006.03.044

[13] A.V. Rao, N.D. Hegde, H. Hirashima, Absorption and desorption of organic liquids in elastic superhydrophobic silica aerogels, J. Colloid. Interf. Sci. 305(1) (2007) 124–132. https://doi.org/10.1016/j.jcis.2006.09.025

[14] C.T. Wang, C.L. Wu, I.C. Chen, Y.H. Huang, Humidity sensors based on silica nanoparticle aerogel thin films, Sensor. Actuat. B 107(1) (2005) 402–410. https://doi.org/10.1016/j.snb.2004.10.034

[15] G.S. Kim, S.H. Hyun, H.H. Park, Synthesis of low dielectric silica aerogel films by ambient drying, J. Am. Ceram. Soc. 84(2) (2001) 453–455. https://doi.org/10.1111/j.1151-2916.2001.tb00677.x

[16] S.W. Park, S.B. Jung, M.G. Kang, H.H. Park, H.C. Kim, Modification of GaAs and copper surface by the formation of SiO_2 aerogel film as an interlayer dielectric, Appl. Surf. Sci. 216(1–4) (2003) 98–105. https://doi.org/10.1016/S0169-4332(03)00488-4

[17] S.B. Jung, S.W. Park, J.K. Yang, H.H. Park, H. Kim, Application of SiO_2 aerogel film for interlayer dielectric on GaAs with a barrier of Si_3N_4, Thin. Solid. Films. 447-448 (2004) 580–585. https://doi.org/10.1016/j.tsf.2003.07.020

[18] J.L. Rousset, A. Boukenter, B. Champagnon, Granular structure and fractal domains of silica aerogels, J. Phys. Condens. Mat. 2(42) (1990) 8445–8455. https://doi.org/10.1088/0953-8984/2/42/021

[19] A. Sayari, A. Ghorbel, G.M. Pajonk, S.J. Teichner, Kinetics of the catalytic transformation of isobutene into methacrylonitrile with NO on supported nickel oxide aerogel, React. Kinet. Catal. L. 15(4) (1981) 459–465. https://doi.org/10.1007/BF02074150

[20] G.M. Pajonk, T. Manzalji, Synthesis of acrylonitrile from propylene and nitric oxide mixtures on PbO_2-ZrO_2 aerogel catalysts, Catal. Lett. 21(3-4) (1993) 361–369. https://doi.org/10.1007/BF00769488

[21] H.D. Gesser, P.C. Goswami, Aerogels and related porous materials, Chem. Rev. 89(4) (1989) 765–788. https://doi.org/10.1021/cr00094a003

[22] I. Smirnova, S. Suttiruengwong, W. Arlt, Feasibility study of hydrophilic and hydrophobic silica aerogels as drug delivery systems, J. Non-Cryst. Solids 350 (2004a) 54–60. https://doi.org/10.1016/j.jnoncrysol.2004.06.031

[23] I. Smirnova, S. Suttiruengwong, M. Seiler, W. Arlt, Dissolution rate enhancement by adsorption of poorly soluble drugs on hydrophilic silica aerogels, Pharm. Dev. Technol. 9(4) (2004b) 443–452. https://doi.org/10.1081/PDT-200035804

[24] T. W. Hamann, A.B.F. Martinson, J.W. Elam, M.J. Pellin, J.T. Hupp, Atomic layer deposition of TiO_2 on aerogel templates: new photoanodes for dye-sensitized solar cells, J. Phys. Chem. C 112(27) (2008) 10303–10307. https://doi.org/10.1021/jp802216p

[25] N.V. Steere, Handbook of laboratory safety, CRC, Boca Raton, Fla, 1971.

[26] F. Schwertfeger, A. Zimmermann, G. Frisch, C. Corp, Use of aerogels in agriculture, U.S. Patent 7,674,476 (2010).

[27] J.E. George, J.M. Pound, R.B. Davey, Chemical control of ticks on cattle and the resistance of these parasites to acaricides, Parasitology 129 (2004) 353-366. https://doi.org/10.1017/S0031182003004682

[28] P. Willadsen, Tick control: thoughts on a research agenda, Vet. Parasitol. 138 (2006) 161–168. https://doi.org/10.1016/j.vetpar.2006.01.050

[29] R.Z. Abbas, M.A. Zaman, D.D. Colwell, J. Gilleard, Z. Iqbal, Acaricide resistance in cattle ticks and approaches to its management: the state of play, Vet. Parasitol. 203 (2014) 6–20. https://doi.org/10.1016/j.vetpar.2014.03.006

[30] S.A. Allan, L.A. Patrican, Susceptibility of immature *Ixodes scapularis* (Acari: Ixodidae) to desiccants and insecticidal soap, Exp. Appl. Acarol. 18 (1994) 691-702. https://doi.org/10.1007/bf00051536

[31] A.T. Showler, W.L.A. Osbrink, E. Munoz, R.M. Caesar, V. Abrigo, Lethal effects of silica gel-based CimeXa and kaolin-based Surround dusts against ixodid (Acari: Ixodidae) eggs, larvae, and nymphs, J. Med. Entomol. 56 (2019) 215–221. doi:10.1093/jme/tjy152

[32] I.B. Tarshis, Laboratory and field studies with sorptive dusts for the control of arthropods affecting man and animal, Exp. Parasitol. 11(1) (1961) 10-33. https://doi.org/10.1016/0014-4894(61)90003-0

[33] A.T. Showler, J.L. Harlien, Effects of Silica-Based CimeXa and Drione Dusts Against Lone Star Tick (Ixodida: Ixodidae) on Cattle, J. Med. Entomol. 57(2) (2020) 485-492. https://doi.org/10.1093/jme/tjz180

[34] P. Golob, Current status and future perspectives for inert dusts for control of stored product insects, J. Stored Prod. Res. 33 (1997) 69-79. https://doi.org/10.1016/S0022-474X(96)00031-8

[35] D.M. Armitage, D.A. Collins, D.A. Cook, J. Bell, The efficacy of silicaceous dust alternatives to organophosphorus compounds for the control of storage mites, Proc. 7th Int. Work. Conf. Stored Prod. Prot. (1998).

[36] J. Schulz, J. Berk, J. Suhl, L. Schrader, S. Kaufhold, I. Mewis, C. Ulrichs, Characterization, mode of action, and efficacy of twelve silica-based acaricides against poultry red mite (*Dermanyssus gallinae*) in vitro, Parasitol. Res. 113(9) (2014) 3167-3175. https://doi.org/10.1007/s00436-014-3978-6

[37] I.B. Tarshis, Use of sorptive dusts on fleas, Calif. Agr. 13(3) (1959b) 13-14. doi: 10.3733/ca.v013n03p13

[38] I.B. Tarshis, M.R. Dunn, Control of the brown dog tick, Calif. Agr. 13(10) (1959) 11-16. doi: 10.3733/ca.v013n10p11

[39] I.B. Tarshis, Control of the snake mite (*Ophionyssus natricis*), other mites and certain insects with the sorptive dust, SG 67, J. Econ. Entomol. 53 (1960) 903-908. https://doi.org/10.1093/jee/53.5.903

[40] W. Ebeling, Control of the tropical rat mite, J. Econ. Entomol. 53 (1960) 475-476. https://doi.org/10.1093/jee/53.3.475

[41] J.F. Odeneal, Silica powder insecticide with glycols to reduce dustiness, U.S. Patent 3,235,451 (1966).

[42] P. Alexander, J.A. Kitchener, H.V.A. Briscoe, Inert dust insecticides: Part 1. Mechanism of action, Ann. Appl. Biol. 31 (1944) 143-159. https://doi.org/10.1111/j.1744-7348.1944.tb06225.x

[43] W. Ebeling, Physicochemical mechanisms for the removal of insect wax by means of finely divided powders, Hilgardia 30 (1961) 531-564. https://doi.org/10.3733/hilg.v30n18p531

[44] W. Ebeling, Sorptive dusts for pest control, Annu. Rev. Entomol. 16 (1971) 123-158. https://doi.org/10.1146/annurev.en.16.010171.001011

[45] C.L. Hockenyos, Effects of dusts on the oriental roach, J. Econ. Entomol. 26 (1933) 792-794. https://doi.org/10.1093/jee/26.4.792

[46] S.F. Chiu, Toxicity studies of so-called "inert" materials with the bean weevil, *Acanthosceles obtectus* (Say), J. Econ. Entomol. 32 (1939) 240-248. https://doi.org/10.1093/jee/32.2.240

[47] H.V.A. Briscoe, Some new properties of inorganic dusts, J. Roy. Soc. Arts 91 (1943) 593-607.

[48] M.K. Krishnakumari, S.K. Majumder, Modes of action of active carbon and clay on *Tribolium castaneum* (Hbst.), Nature 193(4822) (1962) 1310-1311. https://doi.org/10.1038/1931310a0

[49] M.R.G.K. Nair, Structure of waterproofing epicuticular layers in insects in relation to inert dust action, Indian J. Entomol. 10 (1957) 37-49.

[50] W. Ebeling, R.E. Wagner, Rapid desiccation of drywood termites with inert sorptive dusts and other substances, J. Econ. Entomol. 52 (1959) 190-207. https://doi.org/10.1093/jee/52.2.190

[51] R.P. Patel, N.S. Purohit, A.M. Suthar, An overview of silica aerogels, Int. J. Chem Tech Res. 1(4) (2009) 1052-1057.

[52] Anonymous, Farm chemicals handbook, Meister, Willoughby, Ohio, 1982.

[53] A. McLaughlin, Laboratory trials on desiccant dust insecticides, in: E. Highley E.J. Wright, H.J. Banks, B.R. Champ (Eds.) Proc. 6th International Working Conference on Stored-Products Protection, Canberra, Australia, 1994, pp. 638-645.

[54] A.H. Kamel, E.Z. Fam T.M. Ezzat, Silica aerogels as grain protectants, Bull. Soc. Entomol. Egypt 48 (1964) 37-47.

[55] L.M. Redlinger, H. Womack, Evaluation of four inert dusts for the protection of shelled corn in Georgia from insect attack, USDA-ARS (1966) 51-57.

[56] D.W. LaHue, C.C. Fifield, Evaluation of four inert dusts on wheat as protectants against insects in small bins, USDA-ARS Marketing Research Report 780, 1967.

[57] D.W. LaHue, Evaluation of malathion, diazinon, a silica aerogel and a diatomaceous earth as protectants on wheat against lesser grain borer attack in small bins, USDA-ARS Marketing Research Report 860, 1970. https://doi.org/10.5962/bhl.title.63320

[58] F.K. Hsieh, S.S. Kao, W.G. Chen, Tests on control of the maize weevil, *Sitophilus zeamais* Motschulsky, by nontoxic materials, Plant Prot. Bull. Taiwan 20 (1978) 8-15.

[59] V.B. Wigglesworth, Transpiration through the cuticle of insects, J. Exp. Biol. 21 (1945) 97-114.

[60] L.O. Young, Silica flatting agent and a method of manufacturing it, U.S. Patent No. 2,625,492, (1953).

[61] W.A. Donahue, A.T. Showler, M.W. Donahue, B.E. Vinson, W.L.A. Osbrink, Knockdown and lethal effects of eight commercial nonconventional and two pyrethroid insecticides against moderately permethrin-resistant adult bed bugs, *Cimex lectularius* (L.) (Hemiptera: Cimicidae), Biopestic. Int. 11 (2016) 108–117.

[62] W. Ebeling, D.A. Reierson, R.J. Pence, M.S. Viray, Silica aerogel and boric acid against cockroaches: external and internal action, Pestic. Biochem. Phys. 5(1) (1975) 81-89. https://doi.org/10.1016/0048-3575(75)90047-4

[63] R.E. Wagner, W. Ebeling, Lethality of inert dust materials to *Kalotermes minor* Hagen and their role as preventives in structural pest control, J. Econ. Entomol. 52 (1959) 208-212. https://doi.org/10.1093/jee/52.2.208

[64] I.B. Tarshis, Sorptive dusts on cockroaches, Calif. Agr. 13(2) (1959a) 3-5.

[65] I.B. Tarshis, UCLA tests with desiccant dusts for roach control, Pest Control 27 (1959b) 14-28.

[66] I.B. Tarshis, The use of the silica aerogel insecticides, Dri-Die 67 and Drione, in new and existing structures for the prevention and control of cockroaches, Lab. Anim. Care 14 (1964) 167-184.

[67] I.B. Tarshis, Silica aerogel insecticides for the prevention and control of arthropods of medical and veterinary importance, Angew. Parasitol. 8(4) (1967) 210-237.

[68] D.W. Micks, Susceptibility of mosquitoes to silica gel insecticides, J. Econ. Entomol. 53(5) (1960) 915-918. https://doi.org/10.1093/jee/53.5.915

[69] R.C. Moore, Boric acid-silica dusts for control of German cockroaches, J. Econ. Entomol. 65(2) (1972) 458-461. https://doi.org/10.1093/jee/65.2.458

[70] S.R. Loschiavo, Availability of food as a factor in effectiveness of a silica aerogel against the merchant grain beetle (Coleoptera: Cucujidae), J. Econ. Entomol. 81(4) (1988a) 1237-1240. https://doi.org/10.1093/jee/81.4.1237

[71] S.R. Loschiavo, Safe method of using silica aerogel to control stored-product beetles in dwellings, J. Econ. Entomol. 81(4) (1988b) 1231-1236. https://doi.org/10.1093/jee/81.4.1231

[72] N.D.G. White, S.R. Loschiavo, Factors Affecting Survival of the Merchant Grain Beetle (Coleoptera: Cucujidae) and the Confused Flour Beetle (Coleoptera: Tenebrionidae) Exposed to Silica Aerogel, J. Econ. Entomol. 82(3) (1989) 960-969. https://doi.org/10.1093/jee/82.3.960

[73] A. Barbosa, P. Golob, N. Jenkins, Silica aerogels as alternative protectants of maize against *Prostephanus truncatus* (Horn) (Coleoptera: Bostrichidae), In Proceedings of the 6th Int. Working Conf. On Stored-product Protection, Canberra 2 (1994) 623-627.

[74] M.M. Sabbour, Entomotoxicity assay of nanoparticle 4-(silica gel Cab-O-Sil-750, silica gel Cab-O-Sil-500) against *Sitophilus oryzae* under laboratory and store conditions in Egypt, Specialty J. Biol. Sci. 1(2) (2015) 67-74.

[75] J.F. Anderson, R.S. Cowles, Susceptibility of *Cimex lectularius* (Hemiptera: Cimicidae) to pyrethroid insecticides and to insecticidal dusts with or without pyrethroid insecticides, J. Econ. Entomol. 105 (2012) 1789–1795. https://doi.org/10.1603/EC12089

[76] J.M. Desmarchelier, J.C. Dines, Dryacide® treatment of stored wheat: its efficacy against insects, and after processing, Aust. J. Exp. Agr. 27 (1987) 309-312. http://hdl.handle.net/102.100.100/270624?index=1

[77] Y.N. Aldryhim, Efficacy of the amorphous silica dust, Dryacide, against *Tribolium confusum* Duv. And *Sitophilus granarius* (L.) (Coleoptera: Tenebrionidae and Curculionidae), J. Stored Prod. Res. 26(4) (1990) 207-210. https://doi.org/10.1016/0022-474X(90)90023-L

[78] Y.N. Aldryhim, Combination of classes of wheat and environmental factors affecting the efficacy of amorphous silica dust, Dryacide®, against *Rhyzopertha dominica* (F.), J. Stored Prod. Res. 29 (1993) 271-275. https://doi.org/10.1016/0022-474X(93)90010-2

[79] G.N.J. Le-Patourel, J. Singh, Toxicity of amorphous silicas and silica-pyrethroid mixtures to *Tribolium castaneum* (Herbst) (Coleoptera: Tenebrionidae), J. Stored Prod. Res. 20(4) (1984) 183-190. https://doi.org/10.1016/0022-474X(84)90002-X

[80] C.H. Vrba, H.P. Arai, M. Nosal, The effect of silica aerogel on the mortality of *Tribolium confusum* (Duval) as a function of exposure time and food deprivation, Can. J. Zool. 61(7) (1983) 1481-1486. https://doi.org/10.1139/z83-199

[81] Y. Akhtar, M.B. Isman, Horizontal transfer of diatomaceous earth and botanical insecticides in the common bed bug, *Cimex lectularius* L.; Hemiptera: Cimicidae, PLoS One 8(9) (2013) e75626. https://doi.org/10.1371/journal.pone.0075626

[82] J. Goddard, K. Mascheck, Laboratory assays with various insecticides against bed bugs taken from a poultry house in Mississippi, Midsouth. Entomologist. 8 (2015) 10–15.

[83] G.D. White, W.L. Berndt, J.H. Schesser, C.C. Fifield, Evaluation of four inert dusts for the protection of stored wheat in Kansas from insect attack, USDA Agricultural Research Service Report, ARS (1966) 51-80.

[84] G.N. Patourel, The effect of grain moisture content on the toxicity of a sorptive silica dust to four species of grain beetle, J. Stored Prod. Res. 22 (1986) 63-69. https://doi.org/10.1016/0022-474X(86)90020-2

[85] F.L. Watters, Protection of packaged food from insect infestation by the use of silica gel, J. Econ. Entomol. 59 (1966) 146-149. https://doi.org/10.1093/jee/59.1.146

[86] S.L. Gowers, G.N.J. Le-Patourel, Toxicity of deposits of an amorphous silica dust on different surfaces and their pick-up by *Sitophilus granarius* (L.) (Coleoptera: Curculionidae), J. Stored Prod. Res. 20(1) (1984) 25-29. https://doi.org/10.1016/0022-474X(84)90032-8

[87] J. Kane, Silica based dusts for the control of insects infesting dried fish, J. Stored Prod. Res. 2 (1967) 251-255. https://doi.org/10.1016/0022-474X(67)90073-2

[88] I.B. Tarshis, How to apply sorptive dusts for cockroach control, Pest Control 27 (1959d) 30-27.

Keyword Index

About the Editors

Dr. Inamuddin is working as Assistant Professor at the Department of Applied Chemistry, Aligarh Muslim University, Aligarh, India. He obtained Master of Science degree in Organic Chemistry from Chaudhary Charan Singh (CCS) University, Meerut, India, in 2002. He received his Master of Philosophy and Doctor of Philosophy degrees in Applied Chemistry from Aligarh Muslim University (AMU), India, in 2004 and 2007, respectively. He has extensive research experience in multidisciplinary fields of Analytical Chemistry, Materials Chemistry, and Electrochemistry and, more specifically, Renewable Energy and Environment. He has worked on different research projects as project fellow and senior research fellow funded by University Grants Commission (UGC), Government of India, and Council of Scientific and Industrial Research (CSIR), Government of India. He has received Fast Track Young Scientist Award from the Department of Science and Technology, India, to work in the area of bending actuators and artificial muscles. He has completed four major research projects sanctioned by University Grant Commission, Department of Science and Technology, Council of Scientific and Industrial Research, and Council of Science and Technology, India. He has published 178 research articles in international journals of repute and nineteen book chapters in knowledge-based book editions published by renowned international publishers. He has published 120 edited books with Springer (U.K.), Elsevier, Nova Science Publishers, Inc. (U.S.A.), CRC Press Taylor & Francis Asia Pacific, Trans Tech Publications Ltd. (Switzerland), IntechOpen Limited (U.K.), Wiley-Scrivener, (U.S.A.) and Materials Research Forum LLC (U.S.A). He is a member of various journals' editorial boards. He is also serving as Associate Editor for journals (Environmental Chemistry Letter, Applied Water Science and Euro-Mediterranean Journal for Environmental Integration, Springer-Nature), Frontiers Section Editor (Current Analytical Chemistry, Bentham Science Publishers), Editorial Board Member (Scientific Reports-Nature), Editor (Eurasian Journal of Analytical Chemistry), and Review Editor (Frontiers in Chemistry, Frontiers, U.K.) He is also guest-editing various special thematic special issues to the journals of Elsevier, Bentham Science Publishers, and John Wiley & Sons, Inc. He has attended as well as chaired sessions in various international and national conferences. He has worked as a Postdoctoral Fellow, leading a research team at the Creative Research Initiative Center for Bio-Artificial Muscle, Hanyang University, South Korea, in the field of renewable energy, especially biofuel cells. He has also worked as a Postdoctoral Fellow at the Center of Research Excellence in Renewable Energy, King Fahd University of Petroleum and Minerals, Saudi Arabia, in the field of polymer electrolyte membrane fuel cells and computational fluid dynamics of polymer electrolyte membrane fuel cells. He is a life member of the Journal of the Indian

Chemical Society. His research interest includes ion exchange materials, a sensor for heavy metal ions, biofuel cells, supercapacitors and bending actuators.

Dr. Rizwana Mobin is working as Assistant Professor in the Department of Industrial Chemistry, Govt. College for Women, Cluster University, Srinagar, India. She received her B.Sc. Hons., Masters and Ph.D (Applied Chemistry) from Aligarh Muslim University, Aligarh, India on the topic "Studies on Thin-Layer Chromatographic Analysis of Surfactants". She has been the recipient of the Gold medal at Masters level. She has published several research articles in international journals of repute and six book chapters in knowledge-based book editions published by renowned international publishers. She has edited books with Materials Research Forum LLC, U.S.A. Her research expertise includes thin-layer chromatography, development of new methodologies involving green solvent system for the analysis of surfactants and food dyes.

Dr. Mohd Imran Ahamed received his Ph.D degree on the topic "Synthesis and characterization of inorganic-organic composite heavy metals selective cation-exchangers and their analytical applications", from Aligarh Muslim University, Aligarh, India in 2019. He has published several research and review articles in the journals of international recognition. Springer (U.K.), Elsevier, CRC Press Taylor & Francis Asia Pacific and Materials Research Forum LLC (U.S.A). He has completed his B.Sc. (Hons) Chemistry from Aligarh Muslim University, Aligarh, India, and M.Sc. (Organic Chemistry) from Dr. Bhimrao Ambedkar University, Agra, India. He has co-edited more than 20 books with Springer (U.K.), Elsevier, CRC Press Taylor & Francis Asia Pacific, Materials Research Forum LLC (U.S.A) and Wiley-Scrivener, (U.S.A.). His research work includes ion-exchange chromatography, wastewater treatment, and analysis, bending actuator and electrospinning.

Dr. Tariq Altalhi joined Department of Chemistry at Taif University, Saudi Arabia as Assistant Professor in 2014. He received his doctorate degree from the University of Adelaide, Australia in the year 2014 with Dean's Commendation for Doctoral Thesis Excellence. He was promoted to the position of the head of the Chemistry Department at Taif university in 2017 and Vice Dean of Science college in 2019 till now. His group is involved in fundamental multidisciplinary research in nanomaterial synthesis and engineering, characterization, and their application in molecular separation, desalination, membrane systems, drug delivery, and biosensing. In 2015, one of his work was nominated for the Green Tech awards from Germany, Europ's largest environmental and business prize, amongst top 10 entries. His interest lies in developing advanced chemistry-based solutions for solid and liquid municipal (both organic and inorganic) waste management. In this direction, he focuses on the transformation of solid organic waste to valuable nanomaterials & economic nanostructure. Specifically, his research work focuses on the conversion of plastic bags to carbon nanotubes, fly ash to efficient adsorbent material, etc. Another stream of interests looks at natural extracts and their application in generation of value- added products such as nanomaterials, incense, etc. Through his work as an independent researcher, he has gathered strong management and mentoring skills to run a group of multidisciplinary researchers of various fields including chemistry, materials science, biology, and pharmaceutical science. His publications show that he has developed a wide network of national and international researchers who are leaders in their respective fields. In addition, he has established key contacts with major industries in Kingdom of Saudi Arabia.

CPSIA information can be obtained
at www.ICGtesting.com
Printed in the USA
BVHW051357080321
601998BV00011BA/1125